METRO SAPIENS: SMART CITY

PROF: R V M CHOKKALINGAM

Copyright © Prof: R V M Chokkalingam
All Rights Reserved.

ISBN 978-1-63832-233-7

This book has been published with all efforts taken to make the material error-free after the consent of the author. However, the author and the publisher do not assume and hereby disclaim any liability to any party for any loss, damage, or disruption caused by errors or omissions, whether such errors or omissions result from negligence, accident, or any other cause.

While every effort has been made to avoid any mistake or omission, this publication is being sold on the condition and understanding that neither the author nor the publishers or printers would be liable in any manner to any person by reason of any mistake or omission in this publication or for any action taken or omitted to be taken or advice rendered or accepted on the basis of this work. For any defect in printing or binding the publishers will be liable only to replace the defective copy by another copy of this work then available.

The book is dedicated to my late grandmother R Sivabhushanammal w/o R. Venkatachalam Chettiar- founded Sri Subramanyaswamy Temple @ KUPPAM- 77 years ago on 06.03.1944; also dedicated to our Yogini Ramanamma garu of Iruvaram Konda Guha, Chittoor.

Contents

About The Book	vii
1. Urban Sapiens	1
2. Urban Sprawl	17
3. Digital Urbanism	31
4. Smart City	47
5. New Urbanism	63
6. Urban Science	71
Author Bio	83

About The Book

Human Species is now predominantly an urban-dwelling creature- called Urban Sapiens. Cities are arguably the most complex human creation with built in environments, social networks, modified ecologies, economic systems, and political entities. Cities have become an organic and necessary part of human civilization. A fundamental aim of sustainability is to preserve the existence of our species and so we embrace urbanism as a human quality to achieve that end. The author takes the reader through this definitive source book into the lives of Urban Sapiens, and provides a thick description on urban spaces. The tragic landscape of highway strips, parking lots, housing tracts, shopping malls, junked city centres, and ravaged countryside are an expression of our economic predicaments. The timely book explores the range of theoretical and empirical factors that shape the characteristics of Urban Sprawl. The open book draws together cross-disciplinary overview of Digital Urbanism characterized by the key phrases of data-driven society, and knowledge-intensive society. The author explores the fundamental convergence of new digital technologies, tools, and devices in the current narrative of Smart City development. The book highlights that cities must determine to make them a place for Smart Citizens to not only want to live in, but to improve and reap the value of what makes the place special. The author argues by looking into the future, and sees the potential that smart and digital tools, technologies, and optimistic mindsets can play to create cities that enable equitable, financial, social, and cultural prosperity for all citizens. The insightful book

ABOUT THE BOOK

draws out timely narratives and insights into New Urbanism, whose comprehensive urban design and planning seek to foster place identity, sense of community, and environmental sustainability. Metro-Sapiens are city-dwellers of the future, who need to create a symbiotic relationship between the city and the environment with inherent dependencies. The book is a small part of the city in the 21stcentury. The book provides a valuable resource that would enable the reader to take his own critical position within the topic.

CHAPTER ONE

Urban Sapiens

"Parking is a narcotic and ought to be a controlled substance. It is addictive and one can never have enough"- Victor Dover

Urban Sapiens tends to embrace urbanism as a human quality with a fundamental aim of sustainability to preserve the existence of our species. Homo Sapien is now predominantly an urban-dwelling creature. The root 'home' means man, and the root 'sapien' means being- so human being. A modern day human, which is also a homo sapien-is a human being. Homo Sapien is the scientific name for the fleshy, pink, and carbon based creature that currently has the most dominance on the planet known as Earth. All people today are classified as Homo Sapiens. Humans, the dominant species in the globe have shaped the world to suit its own needs. A major step in our evolution occurred when we first began to organize and settle in cities. Cities have been around for thousands of years, since the first were settled in Mesopotamia between 4000-3000 BC. But only over the last several centuries have humans moved into cities en masse. Now, more than half of the world's population can be found in urban areas. Cities are very much the dominant habitat of our species. By 2030, more than five billion people, that is six out of ten human beings will live in cities and urban centres.

Urban Sapiens move towards more sustainable urbanism and their relationship to urbanism is expected to continue with inherent interdependencies. Current population can no longer survive in solely rural lifestyles, and need to embrace their inner urbanites and become Urban Sapiens. The characteristics of urban life are not just density but intensity, not just quantity but quality, and not just location but connectivity. This astounding increase in density in our central place has created a demand for how citizens interact with services, transportation, and infrastructure within cities. The cultural, political, and economic divides are certainly play out in and on our cities via online, and offline, publicly, and privately. The growing global popularity of urban living, makes cities face challenges of quality living. We struggle to contain, plan, and organize capitalist urbanization so as to facilitate to tie the question of urban life to environmental determinism. The idea to move into cities with the right planning could be a strategy toward a more sustainable future. We have to plan for the complexity, and density of our cities, as we experience the higher levels of disruption, which can be stressful, noisy, and dusty.

Urban Sapiens has slim grounds for identifying the key factors possessed by the human brain that accounts for their unique reasoning abilities. Cities are one of the greatest inventions as they are the engines of innovation, providing the density, diversity, interaction, and networks to make us more creative and productive. We are distracted by the lure of shiny, digital, smart technology, and forget the human story anchored in the narrative of city making. Cities must determine to make them a place for smart citizens to not only want to live in, but to improve and reap the value of what makes the place special. We have to fuel the Internet of Things innovation, and turn a beloved community hub into a smart living lab. Cities are now very much the dominant habitat for our species, and we move towards more sustainable urbanism. There is more variability in the social organization of human societies. New situations, and problems arise and make population to change their behaviour. We are becoming city-dwelling Metro Sapiens to achieve sustainability. The term' Metro Sapiens' refers and characterizes city-dwellers of the future.

Urban Sapiens or Homo Sapiens as human is a diurnal creature designed to sleep at night, and to be up and about during daylight. Our longer lower extremely associated with our upright, bipedal locomotion. Like all humanoids, we have a relatively short trunk, long arms, and slender fingers. As primates, we humans are uniquely cosmopolitan, and we are also the most terrestrial of all primates. Human societies are generally multilevel structures of small subsistence units grouped into larger communities. There are many implications of this massive urban growth on biodiversity, ecosystem services, and conservation. Expansion of urban areas is on average twice as fast as urban population with significant consequences for greenhouse gas emissions and climate change. Urbanization implies an increase in job types that are less active than rural ones, this being favourable for the development of problems such as excess weight and obesity. The promise of jobs and prosperity, among other factors, pulls people to cities. The cost of living in urban areas is very high. When this is combined with random and unexpected growth as well as unemployment, there is the spread of unlawful resident settlements represented by squatters.

Urban Sapiens life encourages sedentary lifestyles and consumption of food with few nutritional benefits leading to chronic diseases. Urbanization usually occurs when people move from villages to settle in cities in hope of gaining a better standard of living. It is driven by pull factors that attract people to urban areas and push factors that drive people away from the countryside. Employment opportunities, educational institutions, and urban lifestyles are the main pull factors. Poor living conditions, lack of opportunities, and poor health care are the main push factors. Cities also have advanced communication and transportation networks. Many low-income families gravitate to informal settlements or urban slums that proliferate in and around cities. The provision of water and sanitation services to growing urban settlements, peri-urban and slum areas present critical challenges. Solid waste management in cities is inefficient, which means the proper collection, transfer, recycling, and disposal of all throw away materials. Urban waste often ends up in illegal dumps on streets, open spaces, waste lands, drains or rivers- a problem in peri-urban areas.

Urban Sapiens sees the air quality in cities frequently very poor as a result of air pollution from many different sources such as vehicle exhausts, smoke from domestic fires, diesel-powered generators, dust from construction works, and output from factory chimneys. The combustion of solid waste creates yet another environmental problem. Many cities across the world have grown rapidly over the past 50 years in terms of total population. By 2050, more than two thirds of the world's population will live in urban areas or close to 7 billion people are projected to live in urban areas. We see a strong relationship between urbanization and income: as countries get richer, they tend to become more urbanized; and urban population tend to have higher living standards; but agriculture employment falls with urbanization. Japan's capital city, Tokyo, has the largest population of the world's capital cities at around 37 million people. Urban planning decisions and strategic design thinking in the context of rapid urbanization account for social equity, mobility patterns, global competitiveness, and energy efficiency. The growth of cities in 21stcentury is driving the most significant economic transformation in history.

Urban Sapiens feels that the technology of communication and transportation has made it easier for them to stay in touch with their origin communities. There is no strict dividing line between rural and urban; rather, there is a continuum where one bleeds into the other. While the population of urban dwellers is continuing to rise, sources of social strain are rising along with it. Another driver of urbanization is the shape of modern economies, specially the increasing concentration of wealth creation, and the specialized nature of modern workforce. When wealth is concentrated in smaller groups, and more occupations revolve around products and services that support these centres of wealth creation, there is natural pull towards those centres. Concentrated wealth creation attracts people, which increases creation potential in a virtuous cycle that accelerates urban growth. Urban markets are different from rural or suburban markets for companies who build and sell goods. Urban planners face major challenges to accommodate sustainable growth, and incorporate new technologies in an appropriate, cost-efficient, and environmentally sustainable fashion.

Urban Sapiens views that urbanization has raised the standard of living for many by offering employment opportunities and better public services. Sustainable urbanization can cover all actions that lead to a more efficient management of land and resources, improved mobility, economic dynamism, higher environmental quality, safety, security, access to urban services, and social cohesion. Science, technology, and innovation are key elements of sustainable urbanization, and play a growing role as such. Their use in the urban context implies the application of both high and low technology, and innovative approaches to urban planning. Each urban setting faces different challenges and has different technology needs. A socio-technical system is the way humans and technologies work together to produce outcomes. More positive outcomes may be achieved if cities are designed with an inter-sectoral mindset, for example when addressing energy, water, and public health in an integrated manner. Science and technology can contribute to sustainable urban planning in order to address issues such as urban sprawl, traffic congestion, and natural hazards.

Urban Sapiens uses many types of technology to keep the urban environment civilized habitable and comfortable. New technologies alter the physical possibilities of human settlements, and change the economic, social, cultural, and political relations of everyday urban life. Highways, railroads, street cars, automobiles, and metro-rails lead to successive and dramatic reorganization of urban space. Likewise, electricity, water, sewer, disease control, pollution control, and building construction technologies have had profound impact on urban form and structure. Changes in urban density are significantly affected by changes in urban land area, construction costs, length of all public roads, number of automobile registrations, electricity production capacity, water supply quantity, and mortality rates. Urbanization is a function of technology and society, and the process of technological and social changes are functions of urbanization. Firms have adapted to increased competition for land by building high-rise structures to make more economical use of the expensive land or accessible areas of the downtown. Cities have expanded their areas by annexing land on the fringes of the city.

Urban Sapiens observes city skylines being changed with the evolution of construction technologies, resulting in an intensive use of urban land. High-rise buildings are able to be maintained and operated more efficiently with the increased availability of water and electricity. Certain technologies have helped architects and engineers with the construction of high-rise structures. Besides, emerging technologies- such as information and communication technologies, ubiquitous computing, ubiquitous geographic information, ubiquitous sensor network, radio frequency identification, geo-labelling and sending, global positioning systems, location-based systems, intelligent transport systems etc will help achieve a sustainable global urban living. These technologies promise increased convenience, awareness, transparency, and access to information and social opportunities. They open up new possibilities for addressing urban ills- segregation, poverty, crime, congestion, and pollution. At the same time ITCs and other emerging technologies have been affecting the way modern urban dwellers engage their daily activities. Our policies have to harness the full potential of technologies to make cities better places.

METRO SAPIENS: SMART CITY

Urban Sapiens notices that city population tend to be more diverse in race, economic background, and mindset, which influence office culture and work-styles. With the invention of automobiles, building elevators, smart phones, and so many others have influenced the way cities run and operate. The opportunity for certain products to address the population's needs also contributes to the draw of tech start-ups in these urban settings. Because of the rapid and unplanned nature of urbanization, these trends can aggravate inequality, through slum formation and unregulated expansion. By incorporating technology into urban design, a greater level of connectivity can be achieved for residential and commercial settings. It enables one to work from home and other locations, reducing commute times, and improving traffic conditions. Planning a city with technology helps in the successful design and layout of urban centres, and incorporating it effectively can boost creativity, connections, and trust with a given community. ICT infrastructure is invisible to the public, being composed of underground networks of cables and fibre optics, and satellite-based telecommunications.

Urban Sapiens living culture is reflected in the concept of urban form, and different cities have different cultural backgrounds. City is a kind of the main human living form and living space. The city is a kind of living environment, and people are in the city with a variety of lively activities. The urban physical environment is the space vector of people's lives, so urban planning and urban methods need to be on the perspective of urban culture. Many cities have their historical relics from the surface, and the history of the city has a role in promoting heritage, which is the driving force of urban development. The inheritance and development of historical and cultural traditions together has to meet the new needs of the growing city. It is properly essential to handle these relations conforming to the urban sustainable development, adapt to the needs of social progress, and promote health development of the harmonious urban. Urban planning has to be systematic, dynamic, sustainable, and cultural to meet the needs of contemporary urban culture. Indeed, building on existing social infrastructure to develop cultural and economic activities is an essential element.

Urban Sapiens acts and interacts with one another, and with their physical environment. Urban society is a group of people characterized by a certain way of life, or an urban culture. The urban society is heterogeneous known for its diversity and complexity. The concept of urban society would include a certain system of values, norms, social processes, and relations that in their totality define a historically specific type of social organization. The concept of urban society cannot be separated from both the historical and physical dimensions of the city. Social and economic inequalities in urban areas are also more pronounced than in rural areas, which are made visible by differences in lifestyle. The urban society is mobile and provides more chances for social mobility. Occupations are more specialized and there is a widespread division of labour. Joint families are less, and more than the family, individual is given more importance. People are more class conscious and progressive. Many of the poor are engaged in what is termed as informal employment, which is work that does not have the benefits of registration, permits, and licenses as well as social security required by law.

Urban Sapiens has experienced a consumer evolution at multiple levels. Macroeconomic growth has doubled real incomes and almost all households substantially increased discretionary consumer purchases. Former luxuries such as refrigerators, colour television, and washing machines have become household necessities, now mobile phones. There is the unprecedented upgrading of the quality of urban homes, and the privatization of nearly all residential properties. The elaborate shower fixtures, modular furniture, and illuminated cabinets have become widely available throughout urban cities. Urbanites have enthusiastically consumed globally branded foodstuffs, pop-music videos, and goods. The consumption of different products and services has started to raise at different income levels. The rise of new consumer culture has produced new identities, and hierarchial orders, and configured the social, cultural, and political sphere in which they dwell. Newly built residential and commercial spaces have flourished in cities along with the rising middle class consumers. Consumer aspirations have soared, leading to unrealistically high expectations regarding living standards.

Urban Sapiens lifestyle relates to the way of living to the conditions, and the quality of life in cities. Each city's own activities are supported by the built environment, and by the complex network. The most popular dwellings are no longer single-family houses, but high-rise multiplexes, with high construction costs. Homes are very compact in urban areas with more and more flats and smaller apartments being built. The preference to tiny living or less living space means less time spent on upkeep, less money spent on filling it with possessions, and more of both for all the rest life has to offer. For right now, there is a distinct preference for tiny living in the big cities. But also, increasingly heavy metro traffic and metropolitan sprawl are enough motivation for many to live near their city jobs. Walkability is one of the key draws to their preference for downtown living. City hubs and downtown areas always give off a unique impression and appeal to people who value experiences. The question where to live is based strictly on housing, and distance to work, access to mass transit, and neighbourhood restaurant. Stability and authenticity are increasingly turning into a preference for urban living.

Urban Sapiens tends to feel more expensive to live in urban cities with prices of property, goods, and services going higher. Cities provide for all kinds of resources and goods that cities require, generating the city metabolism. Along with the urban style vibe are the public parks, museums, art centres, clubs, and social activities the offer plenty of options for outgoing people. They have access to good restaurants, popular stores, and theatres. Street fashion is generally associated with youth culture and seen in major urban-inspired stores, and also door-to door online stores. Urban culture and urban clothing goes hand in hand, and the urban style is massive with new urban brands rising. Most fashion bloggers get inspirations from urban ready- to-wear collections. The most common items in the wardrobe of urban wearer are hoodies, t-shirts, and sneakers. All these clothes are usually designed with some sort of slogan. There are jackets, denim pants, and shoes not just for comfort, but as a statement of the urban and democratic lifestyle. Although brands can create garments that can fit more or less in a style, the combination with accessories and way of wearing it is the key.

Urban Sapiens dominates the rural people both economically, and politically. Rural-urban interaction is an important aspect of urbanization. Urbanization has impact on the economy of the surrounding villages. There exists a continuous, or interdependent, or complimentary, or overlapping relationship between rural and urban sectors with mutual exchange systems of goods and services. The urban sector is dependent on rural sector for food supply, for cheap labour, and for vast market for agricultural goods. The urban professionals like doctors, and lawyers, draw a large number of their patients/clients from rural masses because of hospitals/courts concentrated in urban centres. Most rural migrants who move to urban areas are young male occupations, who take up unskilled or semi-skilled occupations. Migration from rural to urban areas has been by translocation, or circulation, or step-migration. There is a difference between artificial environment of urban areas and natural environment of rural areas. Rural work is determined by seasons and weather, whereas urban work is carried out indoors in predictable conditions.

Urban Sapiens visits shopping malls that contain restaurants, theatres, banks, service stations, parking areas, and many retail establishments. Urban shopping plazas are large indoor shopping experiences, usually anchored by department stores. Urban shopping centres have evolved into large-scale, multi-use, lifestyle complexes with their role in the social, and cultural fabric of metropolitan society. Shifting theories of city planning have profoundly altered people's lives everywhere. The questions of fragmented communities, transient population overcrowding, inequality, and segregation are all important issues to be addressed. Cities are inherently sites of conflict and violence. However, they are generally managed through social, cultural, and political mechanisms. The reasons for urban conflict and its potential descend into violence are city density, poverty, inequality, and unemployment. Urban governance is integral to the management and resolution of conflict, and the mitigation of violence. Cities become fragile, when there is a failure to fulfil the core functions, such as ensuring safety for citizens, property, and infrastructure, or access to base services.

Urban Sapiens is sure that entertainment is an essential component of vibrant urban life. Urban entertainment buildings are the most important and iconic, and become a source of pride and a hallmark for the quality of life they offer to urbanites. A well-connected, amenity-rich sports facility or stadium becomes anchor component of many cities. Amusement parks provide thrills, spills, and roller coasters, drawing huge crowds all through the year. Water parks or theme parks attract those who are just looking for a nice time. There are plenty of local live concerts, and simple events on nearly everyday of the week. The Arts galleries have a huge role to play as community glue, where people bump into each other, sharing experiences. All such arenas are destinations that create a footfall, and are therefore great for cities in terms of their longevity and sustainability. Convenience of public transportation, endless entertainment options, quality restaurants, shopping malls, social events, and medical care are the main sources of urban attraction. There is a growing need for properly designed cities to ensure urban stability, and resilience.

Urban Sapiens have chosen to live in a city based on such factors like weather, cost, demographics, proximity to work, public transport, and housing facilities. Cities have an array of options for spending free time in green spaces, theatres, music venues, museums, and other cultural centres. Cities have unlimited social potential for friends and networking with a great diversity in common spaces. Cities are vibrant and thriving centres of culture that offer a wealth of opportunities to engage in and experience the local community. Cities offer local deliveries, online services, and cleanliness standards, and social distancing protocols. Cities present a range of options for dining, shopping, brand stores, and artisan crafts all within a relatively small radius. An excellent perk of living in the city is the sheer amount of education, career, and volunteer opportunities available. Walkers are practitioners of the city, for the city is made to be walked. Cities buzz with energy, where people around doing something, and that can be exciting. Cities could be seen as quintessentially human, an expression of our deep need for social interaction.

Urban Sapiens experiences the cacophony of car horns, screeching brakes, and conversations filling the urban streets- our urban cities have more cars than trees. Trees cool down cities by offering shadow, and releasing water vapour. Trees reduce carbon-dioxide, but they also make people happier, healthy, and more productive- two trees provide enough oxygen to support the family of four. Greater ambient noise and visual stimulation in cities is associated with poor mental health. Nature based solutions for sustainable urbanization relies in large part on natural areas, and features in and around cities to generate essential ecosystem services. The increase of green space may facilitate social cohesion to cooperate with each other in order to survive and prosper Living in a city brings with it access to many amenities and services, that are not frequently available in rural areas. Most inhabitants of urban areas have non-agricultural jobs. Most suburbs are less densely populated than cities, but serve as the residential areas for much of the city's workforce. Suburban migration is to escape the traffic, noise, or to enjoy a larger residence.

Urban Sapiens wonders why so much urban growth is happening? In addition to pull/push factors driving people from rural areas, the positive benefits of being in city with a useful way of structuring society are the core ingredients. The economic benefits of urban form are strong enough that they drive most societies toward urbanization. Economic studies show that it is the extreme potential for interaction that makes cities as centres of productivity, innovation, and creativity. Health studies reveal that the increase in stress and greater prevalence of some mental disorders in cities. Ecological studies indicate that natural features in cities are disintegrated from citizen's lives and green lungs are turning into concrete jungles. Urban growth has directly affected natural habitats through land expansion, while the urban form and consumption patterns have affected greenhouse gas emissions, natural resource use, and water security. One aspect of urban health penalty is obesity and its associated diseases. However, we are living in the urban century, and humanity is in the midst of the greatest migration in our history in absolute terms. Cities are becoming more lucrative.

Urban Sapiens are aware that digital urban planning approaches involve data-driven modelling and visualization offer new ways of understanding how cities function, and predict how changes in their design could affect urban life. They use digital 3D visualization to place models in their real time environment, implement them on a virtual basis, and explore them up-close from every angle. Using 3D city models and speeds up the development of modern-day cities, especially Smart Cities, which makes it easier to secure acceptance for construction projects from stakeholders. A digital is a virtual model of a process, product, or service. Digital Twins- virtual replicas of a physical product, process, or system- bridge physical and digital worlds, by using sensors to collect real-time data about a physical item. This data is then used to create a digital duplicate of the item, allowing it to be understood, analyzed, manipulated, or optimized. Other terms used to describe the technology over the years have included: virtual prototyping, hybrid twin technology, virtual twin, and digital asset management. The concept of digital twin has its roots in engineering, and the creation of engineering drawing.

CHAPTER TWO

Urban Sprawl

"What is bad about sprawl is not its uniformity, but that it is so uniformly bad"- James Kunstler

Urban Sprawl is generally referred to the unrestricted growth in many urban areas over large expanses of land with little concern for urban planning. Urban Sprawl is caused in part by the need to accommodate a rising urban population. As the population of an urban centre increases, its needs for infrastructures such as transportation, water, sewage, and facilities such as housing, commerce, health, schools, and recreation increases, most often resulting in the phenomenon known as Urban Sprawl. Sprawl tends to occur where property values are lower on the periphery of urban centres. Many authorities argue Urban Sprawl diminishes the local character of the community. Urban Sprawl has been correlated with increased energy use, pollution, traffic congestion, a decline in community distinctiveness, and cohesion. However, increased affluence, attractive land, and housing prices, and the desire for larger homes with more amenities play significant roles at the level of the individual. One of the most obvious environmental effects of widespread building construction is the destruction of wildlife habitat.

Urban Sprawl is the rapid expansion of the geographic extent of cities, which is often characterized by low-density residential housing, single-use zoning with increased reliance on the private automobiles for transportation. Many urban planners maintain that modern suburban zoning laws have done much to promote Urban Sprawl. In many cases Urban Sprawl has occurred in areas experiencing population explosion, especially in developing countries. As homes and businesses spread out further apart, the costs of providing community services increases. Continued road-building projects, most notably the onset of the interstate highway, and other infrastructure development makes it possible to build homes on land that was previously inaccessible. Much of the growth in a metropolitan area occurs at the fringes, as large amounts of resources, and services are directed there. Homes built deep within housing tracts are located far away from stores, schools, and employment areas, where residents depend upon automobiles. There is a clear correlation between Urban Sprawl and the epidemic levels of obesity, and increase of chronic diseases associated with physical inactivity.

Urban Sprawl is sometimes caused by equal division between local population increases, and lifestyle choices. Land-value, however, is often considered the chief driver of development patterns. Many experts also believe that weak planning laws and single-use zoning also contribute to Urban Sprawl. Some citizens move to the suburbs to enjoy a lifestyle that is ostensibly closer to nature. Other citizens move to escape the congestion, crime, and noise of the city. Over time, this migration to the suburbs, along with rising local populations, leads to substantial increases in the geographic extent or spatial footprint of metropolitan areas. Construction at the urban fringe is increasingly characterized by a standardization of designs and specifications. Urban Sprawl increase car and truck traffic by creating longer, and more frequent commutes, which leads to a major increase in air-pollution, and ground-level smog. Ubiquitous retail chains with extravagant signage, and facades are often the first to move into newly developed areas. New construction typically occurs on land formerly used for agriculture, as this land is converted to urban use.

Urban Sprawl refers to the migration of a population from populated towns and cities to low density residential development over more and more rural land. Economic prosperity allows many citizens to purchase single-family homes, and private automobiles causing the Urban Sprawl. Exurban low-density neighbourhoods consume more energy per capita than their high-density counterparts closer to the city's core. Suburban residents retain a connection to the city through their automobiles. Massive influxes of new residents contribute to increases in their individual spatial footprints. In some metropolitan areas, relatively modest population growth is also accompanied by significant spatial growth. Some social scientists have linked this Urban Sprawl trend towards design standardization to the rising influence of globalization. Urban Sprawl threatens productive farmland, transforms parks, and open spaces into highways and strip malls. Energy for heating, cooking, lighting, and transportation is largely produced by burning fossil fuels, such as gasoline, home-heating oil, natural gas and coal, a process that contributes to air pollution.

Urban Sprawl increases the physical and environmental footprints of metropolitan areas, which phenomenon leads to the destruction of wildlife habitat, and to the fragmentation of remaining natural areas. Scientists have argued that sprawling urban and suburban development patterns are creating negative impacts including habitat fragmentation, water, and air pollution, as well as increased infrastructure costs, inequality, and social homogeneity. However, in many metropolitan areas, it results from a desire for increased living space, and other residential amenities. Suburban land is relatively inexpensive, and so the homes constructed on this land affords more spac to their occupants than inner-city dwellings. In many areas, urban development pressure, and increased property taxes are forcing farmers out of business. To make a way for human dwellings, and their associated infrastructure, natural land is ploughed under, graded, and paved. The footprint of major metropolitan areas, suburbs, and small towns, ultimately shapes the environmental, and social conditions within the communities. The expanding footprint of cities implies a greater expenditure provision to adjoining fringes.

Urban Sprawl is a phenomenon occurring in several countries, as worldwide people are moving to cities. There are many factors that contribute to Urban Sprawl. Slow-moving streams of suburbs are often channelled to provide more efficient drainages for housing tracts, and commercial areas. People living in sprawling cities have higher living costs, shortage life expectations, increased risk of obesity, and lower economic mobility. The monitoring, and analysis of urban expansion have become a popular topic in geosciences applications in various regions around the world. But, as time goes on, and development sprawls, it is harder to tell where cities end, and suburbs begin. Wild forests, meadows, and wetlands are also appearing, replaced by pavement, buildings, and sterile urban landscaping. The end result is the spreading of a city, and its suburbs over more and more rural areas. It is quite natural that Urban Sprawl could easily lead to some less green space. Urban Sprawl is most pronounced in wide rings of transport corridors. Unprecedented Urban Sprawl at the heart of cities, and communities evokes an urgency, and responsibility to plan and design a better sustainable future.

Urban Sprawl is one of the major outcomes of transformations resulting from population agglomeration in urban centres. When new settlements are developed, neither distance from workplace nor gasoline prices are much mattered in determining the locations of new constructions. Cities will usually have high property taxes, and people usually avoid these taxes by living in the outer suburbs, because the taxes there are usually lower than they would be in other situations. As cities grow, perhaps, our most serious concern has been how they expand out into the surrounding countryside, or sprawling outwards. Middle class select to live in high-density, multi-storey apartments that are starting to spring up around the city, reducing the need to subsidise longer distance, road-based travel by private car. Urbanization is accelerating fast, placing huge pressures on the city core with a huge number of relatively high-rise buildings emerge in the urban core. The urban poor are forcibly relocated into outer-city settlements or located on the urban periphery. Property developers tend to prefer Greenfield developments on the peripheries to the complexities of Brownfield regeneration.

Urban Sprawl is generally typified as low-density, haphazard development spiralling outward from urban centres. Urban sprawl is the result of a complex set of interrelated socio-economic, and cultural forces. City visitors are charmed by the pedestrian streets that thread their way through a maze of buildings, and on narrow streets leading to cafes, where people sit at café tables. There is a striking global trend towards increased street-network sprawl. Urban centres have moderate, and relatively stable levels of street connectivity. Every street-network sprawl predicts the connectivity of later construction at various geographic scales. Streets are a more permanent feature of cities; once laid down, their routes almost never change, even in the face of disasters. Recent urban growth has increasingly resulted in inflexible, and disconnected street-networks. But is connected street routes will remain as a fundamental constraint. Residents move to suburbs in droves in search of more favourable living conditions, where more space, privacy, and affordability offers, what some consider to be more comfortable lifestyle.

Urban Sprawl is a pattern of uncontrolled development around the periphery of a city, which is an increasingly common feature of the built environment around the world. It is a real estate development resulting in low-density, scattered, discontinuous car-dependent construction, usually on the periphery of declining older suburbs and shrinking cit centres. As suburb areas develop, cities expand in geographic size faster than they grow in population. When cities grow, the surrounding land, and the natural green areas are engulfed to build houses, roads, and pathways to match the needs, and desires of the inhabitant population. Urban Sprawl has fundamentally reshaped our urban landscape. There is a growing economic inequality among metros, where people are increasingly sorted into either poorer or richer regions based on their ability to meet their costs. The suburbs on new urban lands add to deterioration of the environment through depletion of resources such as air, water, and soil. Not many of us realize the less obvious, but serious consequences that urban spread can have on the life of the inhabitants.

Urban Sprawl occurs largely because land owners and developers make choices that promote their own economic and personal interests, which do not necessarily coincide with public good. It is low levels of some combination of density, continuity, concentration, clustering, centrality, mixed uses, and proximity in urban area rather than metropolitan region. People who live in large metropolitan areas often find it difficult to travel even short distances without using an automobile. This happens because of the remoteness of residential areas, and inadequate availability of mass transit, walkways, and bike paths. The increased use of motor vehicles releases chemicals, and particles like hydrocarbons, carbon-monoxides, and nitrogen oxides leading to air pollution and smog. There are large volumes of water run- off, because road, and cement do not absorb rain as good as planted areas. Several towns located just beyond the edge of a big city have disappeared, and have been swallowed by the advance of urban spread, and lost their unique identity. Geography often corresponds to denser development because mountains, and deserts around the city prevent urban regions from expanding outwards.

Urban Sprawl is all the more an object of attention if it occurs at a fast pace, and changes environmentally attractive areas into expanses of concrete, and billboards. One particular characterization of spatial form called Urban Sprawl is considered as a contributor to climate change, with environmental consequences from land to water to air. The spread of urban areas outward normally penetrates into rural areas, farmlands, and forests lying on outer edges of city. Peripheral growth can be contagious with existing areas or cut off from them, which can be linear or concentrated. The unplanned development has resulted in overcrowding of services, and infrastructure. The residents of sprawls spend higher portion of their income on transportation than the residents living closer to the city centre. Urban Sprawl can reduce water quality by increasing the amount of surface runoff, which channels oil, and other pollutants into streams, and rivers. The subject of Urban Sprawl is receiving increased attention in both popular media, and scholarly literature. Households struggle to meet a mortgage, and car-loan payments, adults lacking exercises, and teenagers get bored in shopping malls.

Urban Sprawl truly causes havoc in the natural ecological balance or even the lifestyle of the inhabitants. People love to find areas that are less trafficked and more calm, which leads them to sprawl out to other sections of the city. Unprecedented development, cutting trees, loss of green cover, long traffic jams, poor infrastructure force people to move out to new areas. Because of the urban sprawling, population begins to use their cars more often, which means that there is more traffic on the roads more air pollution, and more auto accidents that people have to worry with. Urban Sprawling increases consumption of energy as it is dependent on lengthy distribution systems that undermine efficient energy use. Also the economic effects would be substantial as the costs of public services are much higher in sprawled areas. Sprawling areas are easy to discover on aerial photographs, but it is not always so easy to explore them on maps and data sources. Rapid urbanization causes city centres to experience higher density with high decline in periphery settlement. However, as economic growth continues, people with some wealth, typically the middle class begin to migrate towards suburbans.

Urban Sprawl may involve 'edge cities': low-density residential developments, that give rise to business activity like office buildings, retail, and even manufacturing. Researchers have found that people living in sprawling suburbs spend less time walking, and weigh up more than those living in pedestrian-friendly neighbourhoods. The negative effects of Urban Sprawl are commuting and driver stress, loss of natural environment, loss of social capital, loss of community, and negative health effects. Further, research studies indicate that traffic congestion also has adverse effects on one's blood pressure, mood frustration tolerance, illness frequency, work absences, and job stability. Road range, and aggressive driving are caused by rushed or behind schedule, and increased congestion, and traffic. Walking, and biking are healthier alternatives, but there is usually little emphasis on walking, and biking services in car-culture. Communities are not walkable or bikable, as they need to drive to schools, shops, parks, entertainment, etc, and become more sedentary. Social interactions in Urban Sprawls are minimal as compared to interactions in other places, such as the rural areas, and urban centres.

Urban Sprawl cuts into our precious free time, and contributes to expanding our waistlines. Cities tend to be more compact, and house higher numbers of people in multifamily dwellings. Most places continue to sprawl over time, because either their suburbs are growing faster than their cities, or because they hollow out as their economic fortunes wane. If housing is out of reach, it is easy to see why many may choose suburbs. Perhaps the most basic defining characteristic of cities, and the regions they anchor is how are physically built, but outward, and upward. Every morning, those living in rural areas come to cities using public means for transportation of agricultural produce, which are perishable goods, that must be consumed as soon as they are produced- this contributes to traffic jams. Other people living in rural areas, and working in cities are accused of causing unnecessary accidents since they over speed in rural roads. Destruction of agriculture has led to problems, because more land is covered with impervious material, including concrete. However, the cities, and their suburbs cannot withstand the population growth.

Urban Sprawl has been a predominant way of life for much of the world today. With the present concerns over climate change, oil scarcity, and social isolation, the future of urbanism is a highly debated topic. The loveliest villages at the outer edge of a growing city have a spirit like no other, who dearly care for the community that they serve. There is a vivid ongoing discussion on the variations of planning, urbanization, and migration crisis. In the recent decades cities have experienced constant morphological changes, and demographic transition. Increasing number of low-income people live in suburbs, based on the socio-economic context at the local level. Non-linear expansion waves alternating settlement densification and scattering are found in complex urban context. Urbanization process reflects heterogeneous spatial forms and more socio-economic dynamics. Land degradation has expanded as a result of a variety of factors, including economic, and population growth, land-use changes, and climate variations. Recent urbanization leads to a more evident distinction in historical inner cities, consolidated urban periphery, and peri-urban areas.

Urban Sprawl areas have a growing need for better connections with inter-urban green areas, which are only plant cultivation, but also animal husbandry, fisheries, and any activity occurring along the value chain from production to consumption, is the source of food and sustenance for all cities population. Beyond food, agriculture also provides a critical source of income for the suburban people to live and work at greater distances from city centres and to use land more extensively. Transit villages need to be linked with the mass transit system. The loss of compactness has increased infrastructure costs, and has made agricultural activities more difficult to carry out in the urban fringe. Employers bemoan the loss of time on congested roads, environmentalists condemn the loss of precious natural amenities, and health-care specialists lament the increasing rate of diseases, and the lack of exercise. Driving to work, driving to dinner, driving to meet friends- this is quintessential invention of urban life that is infamously sprawling. Urban Sprawl exists only because it is an outgrowth of activities, or in other words, auto reliance contributes to urban sprawl.

Urban Sprawl is an ill defined, and complex concept, and this contributes to the difficulties in addressing it. There are a variety of proposed solutions to sprawl, the most common being growth management laws, urban containment policies, and zoning restrictions. The use of technology in resolving issues related to urban sprawling is a costly task since it does not offer adequate solutions to social and economic problems. Public funds are spent in infrastructural development that does not benefit the urban dwellers, yet the budget stems from the city council. Apart from consuming land meant for agricultural production, public funds meant for development in rural areas are channelled to the repair of roads, maintenance of water pipes, and provision of additional electricity. Finding a way forward could begin with more meaningful planning and coordination at all levels of government. Strong state-level regulation likewise prevents developers from putting metros against one another in a zero-sum game of attracting constructions. Regional co-ordination, and even intervention at the federal level can deliver the resources and vision necessary to chart a path forward for cities and suburbs.

Urban Sprawl would lead to dramatic negative consequences regarding land cover, local climate, emissions, and pollution, water and ground water, flora, and fauna. The end point of sprawling megacities is total human annihilation, falling victim to technical rationality, and the end of organic human existence. Urban Sprawl is probably best described as the antithesis of an ideal city type- the compact, efficient pedestrian settlement with a rich civic culture. The unique identity of a city has many origins from the cultural and social, to the historic and geographic. Urban areas are still sprawling as they are built out, or their large cities depopulate. People transport themselves, and they have to drive more often, and for longer distances. There is rise of the box store- a form of corporate colonialism, going into distant places, and strip-mining them culturally, and economically. They do not buy from local manufacturers or producers, but larger national firms. Cities characters have been stripped as the unique local businesses have been outmatched. With massive urban growth still to come, is it materially, or ecologically possible to sustain 21stcentury petrol-soaked world.

Urban Sprawl presents one of the most urgent challenges of our times. It undermines the cost-effective provision of public services- roadways, sewage lines, water supply, trash collection, housing facility, community development, police protection, fire services, and public parks. Today/ urban areas generate more than 90 percent of the global economy, are home to more than 50 percent of the world population, consume more than 65 percent of the world energy, and emit 70 percent of global greenhouse gas emissions. It is clear that the diffusion of vehicles as a means of transport is behind the increase in sprawl. The cities grow when the cars begin to occupy their streets and avenues, affecting urban design and the way these cities grow. In relation to mobility, it is assumed that sprawl implies greater distances and lower population density, which hinder the success of mass transit. One of the main characteristics of sprawl is that in which residences, and commercial properties line up on roads extending outward from urban centre, which is called Ribbon Development. The way in which we develop urban landscape, and its areas is a critical component in creating a sustainable city.

Urban Sprawl is not only a physical phenomenon of the dispersion of buildings, and expansion of the space occupied by the city, but also a phenomenon that encompasses different disciplines: geography, urban planning, economics, environmental analysis, sociology, and even policy science. Urban Sprawl has evolved into a exceptionally complex public policy problem. Regional planning programs aim to reduce the spread- low-density, discontinuous, suburban-style development. Planning policies, and techniques can help protect or restore a region's natural resources. Local planning policies can help guide growth in a more ecologically sustainable fashion, and assist local communities in attaining the intended spatial design. And land use intensities. Urban decision-makers, and citizens will need to not only re-connect to nature, but also adopt policies to integrate nature into our daily lives. Sprawl is understood to imply soil predation with consequences on the natural environment of cities, and generates greater distance in the city, which makes mobility more difficult. There are social tensions in cities for different reasons, which are manifested in high rates of crime and violence.

Urban Sprawl is caused by two driving forces, one is exogenous pull power, and the other one is endogenous growth power. A city is a, and permanent human ecosystem, which provides a lot of services, and opportunities to its citizens. Cities allow us to work, create, have fun, and express ourselves together while sharing urban spaces. With the rising costs of providing local public services, and local population growths have brought in fiscal pressures on local governments spending on current account spending, capital outlays, and spending on public safety. Urban infrastructure planning enables people to live, move, and interact faster and safer. Many cities are enhancing quality, and performance of urban services by being digitalized, intelligent, and smarter. The concept of a Smart City has emerged with the help of developing technologies against population growth, increasing urbanization rates, and rapidly depleting resources. Integrating urban and digital planning, Smart Cities are being seen as solutions to the challenges of urbanization, and retainable development. Smart Cities are now arguably the new urban utopias of the 21stcentury.

CHAPTER THREE

Digital Urbanism

"Urbanism works when it creates a journey as desirable as the destination"- Paul Goldberger

Digital Urbanism talks about a kind of city that is substantively an open, complex, and adaptive system based on computer network, and urban information resources, which form a virtual digital space for a city. It is originated by an urban strategy that aims to improve the quality of life for citizens through the agency of new digital technologies connecting different stakeholders and offering a better service. It is such a kind of urban area consisting of the physical elements, the social communities, and the technological infrastructure in the urban context. It is a creation of ICT or Information and Communication Technology service marketplace, and resource deployment. Its objective seems to be that through ICT, citizens participate more in urban decision-making with a multi-stakeholder approach. Collaboration, digetalization, automation, the Internet of Things, and virtualization are some key concepts that come with the continued development of ICT. The hype around digital cities continues to grow and evolve with a number of interconnected solutions that work together cohesively. There is actually no silver bullet or one key technology that cities should embrace.

Digital Urbanism is an emergent understanding of city life shaped by the influx and pervasiveness of digital technologies. Digital technology has always played an important role in shaping cities. Cities have approached their digital transition over the last 15 years- as means of driving change in cities. Cities are taking advantage of digital transition as a goal itself. The Digital City is connected to the net, equipped with technological platforms for information and communication management that can enable the Internet of Things. Urban Life is now significantly shaped by our use of smart phones, embedded sensors, and smart systems that guide our decisions and offer new types of experiences. The properties of digital technology promise great advances and an improved quality of life. The digital revolution is creating a hyper-connected and collaborative society that transforms relationship between people. Across the world we are seeing increased interconnectivity, and cities and people are becoming technology-dependent. Ubiquitous super-fast Internet connectivity has facilitated the digitalization of cities and millions of people. The road to becoming a digital city is not easy.

Digital Urbanism is an attempt to define and understand hybrid spaces as well as ways of experiencing and behaving in our cities. The great thing about a Digital Life is it can be anything we want it to be. New digital technologies that will change our cities are being rapidly deployed. Digital technologies have been transforming our human experiences, and we have seen an increased human-computer symbiosis. Digital Lifestyle makes it easier for people to connect around the world using the Internet and digital technologies. There is no doubt that digital has the power to enhance the quality of our lives, both now and in the future. With the noteworthy growth of Internet of Things and smart devices human life and habits have clearly been changing. Digital World presents many possibilities and benefits to individuals and companies. The use of new digital technologies to communicate and access information is changing the way society works. Digitization can, on the other hand, be seen as a methodology, and refers to making information accessible to both people and computing devices. Citizens live an increasingly Digital Life both in the public and private spheres.

Digital Urbanism promises great advances and an improved quality of life and thereby builds great future cities. Digital Lifestyle is just another form of independence, giving people more flexibility to create the life we imagine. However, the process of digitization of human experiences has been accelerated during the pandemic. In the light of measures like physical distancing and lockdowns, many of us are now moving various aspects of our lives to digital environments. With the recent spikes in infection, people continue heavy smart phone use, and consume more online services. Across the board, consumers are using their tech devices far more than they did before the pandemic, avoiding crowds, and adopting physical distancing. The pandemic appears to be changing the game entirely and removing barriers associated with the habitual use of digital technologies. Many of us have been reluctant to do certain things online, but now the current situation demonstrates that we feel doing them. Going digital is the only way to keep up with the rapid pace of the modern world. Indeed, what is of crucial importance to cities is not what technology is used, but how it is used.

Digital Urbanism has the technology base, which include a critical mass of smart phones and sensors connected by high-speed communication network. Digital Lifestyle means we are always connected-to the Internet, to our devices, and to each other. The planned digitization of many businesses, education institutions, and non-profit organizations has been accelerated. When used correctly, digitalization, and new technologies can be harnessed to transform cities into platform of open innovation, and the city and interact with each other. Consumers today enjoy the convenience of digital service delivery service that many private sector companies now put at their fingertips. There is the global rise of Digital Cities- cities where data-driven decision making positively affects citizen outcomes. However, limited resources place an added burden on city infrastructure and resources as there is population growth, congested streets and highways, and rapid spreading of diseases. The digital transformation of cities is a journey that requires a long-term vision and constant evolution. Digital technologies are used by cities, companies, and the public, and their applications encourage people to adopt and manage to change.

Digital Urbanism is the interplay of multiple interests and actors in the context of the city, as mediated by information and communication technologies. Known online shopping to social media, people can pay bills on our mobile phones. Digital banking for example allows individuals and companies to open an account, access payments, savings, and credit products without ever stepping into a bank branch. This is possible through digitization, which can essentially turn a smart phone into a wallet, a bank branch, and even a library. The Digital Lifestyle enables people to work mobile, anywhere in the world where there is Internet connectivity, for example home, office, hotel, coffee shop, or any place etc. We have access to massive amounts of human knowledge in the time it takes to download a web page. We have wireless broadband, GPS, Internet connected sensors and devices, and an array of applications that change how we interface with the city and interact with each other. Wireless communications have become embedded in our environment and spawned a new kind of a city. Digitally transformed cities keep their residents better informed with real-time updates and serve their residents with services.

Digital Urbanism will provide us with a lens to understand the issues and opportunities, and to develop practices and methods to help build great future cities. Knowledge about tech products and services are now an inherent part of consumer's Digital Life. It is now sometimes impossible really to get away from work, as cell phones with email capacity are functional almost anywhere on the globe. As a result, these digital communication tools make workers more productive, but they also make the employee more connected to the place of work. This makes cell phone users universally available and brings enormous communications capacity right to our pockets or purses. Work hours have increased, and many work harder and longer. The speed and reach of contacts we can make and maintain means that everyone is findable and reachable with a quick search engine query and an email. Faster than anyone really expected, we have become fully immersed in a Digital Lifestyle. There is a profusion of Internet of Things, digital platforms, automation, and sensors across all domains of urban life. Digital Cities will help facilitate data between parties to improve citizens' lives.

Digital Urbanism can make people disconnect physically, and reconnect online. Also many jobs can be done from our home or anywhere. The digital technology has also created now more opportunities for remote employment. A company's workforce has become dispersed as satellite offices and can operate as if they were all centrally located. These workplace changes are restructuring the face of the workforce, and the possibilities for how individual lives are structured as well. There have been corresponding changes in retail, banking, and other industries. Now we know mobile is the future of everything- at least at this moment. Digital Life is about streaming and being more efficient in the way we work. It frees us from repetitive tasks and gives us the opportunity to challenge ourselves, and pursue innovation and creativity. Not only are more people now living in cities than rural areas globally, but more people are connecting to cyberspace through new digital technologies such as smart phones, laptop computers, and other devices. We use new, fast, and frequently changing digital technologies to solve problems. Digitization is a key enabler for dealing with the challenges of cities with new ways.

Digital Urbanism could affect various quality-of-life dimensions, including safety, time, and convenience, health, environmental quality, social connectedness, and civic participation, jobs, and the cost of living. Living an accelerated life will have more stress, and real threats to our freedom and privacy. We are yet to understand the long-term impacts on our society as a whole. We are living in an accelerated pace like on a train running at very high speed that is taking us nobody knows where. The fact that human experiences are more digitized also means that they can in any way, be programmed. It means more controllability of what we are exposed to and the consequences of how we behave in these digital environments. At some moment in the future, the question of who owns or controls the algorithms will become the prime question of humanity. The extension of current trends will lead to a widening economic divide that leaves the majority in the dust of the privileged class. In the future the Internet will make lives both better and worse, providing greater access to information and push horrific misinformation to people.

Digital Urbanism is an asset for cities, not only for business development, and job creation, but also for city governance and getting closer to citizens. The process, supporting societal and urban transition, has a strong impact on governance, and on how our everyday life is organized as well as on the way we make the city work. Beyond the digital divide issue, private data protection and free choice, this trend follows new consumption and production patterns, as well as interaction between people. We will continue to have problems of community and identity online, where malicious actors quite easily pose as others and manipulate people's opinions. The ability of the news media to report facts will be hampered by a cascade of alternate news with different video and audio of the exact same event. We will receive different news, again exacerbated by the prevalence of fake news that is exceedingly difficult to discern from reality. The digital divide will not be one of access but of security, privacy, and autonomy. The design of digital services is increasingly important for how cities are planned, built, and lived in. People expect the same level of service from government as they receive from online retailers.

Digital Urbanism with its tech boom of the past two decades has given rise to large companies like Google,

Amazon, and Facebook whose services are largely independent of geographic location. Our professional and personal life will be tethered to a provider-likely Amazon or Google, which will maintain and run our smart homes, hospitals, schools or offices. Digital devices have simplified many aspects of our lives, but consumers still have concerns about the use of digital technology. Worldwide we already see a rise in authoritarianism a weakening of democracy and the dominance of transnational corporations. Concerns about data security and hacking into digital devices also remain high. Photoshop, after effect, are such tools which can manipulate the original data to something new creating confusion. It is not that digital technology will be a net negative. Given that there is enough technology available, there are a number of bad actors on the human stage to do destructive things. Life will likely involve wearable and ubiquitous computing based on Internet and plat-formed communication. These kinds of tools will likely be available only to those with economic and cultural capital to access them.

Digital Urbanism has the rhythms of fast-paced city life bombarding us with visual stimuli from street signs and posters, through LED displays rushing past, to the personal digital devices on our desk in our hands making our lives as animated corporeal existence. There will be many opportunities for consumers and entrepreneurs in the Internet of the future. People will be mostly better off in 50 years time largely because of our ability to apply things we already know. If the political power in control of knowledge is benevolent and progressive the future will lead to positive societal changes. The main challenge is whether or not we have the social, political, and educational imagination to adapt and effectively use these technologies.

Again, we continue to forget that 75 percent of the world's populations are effectively peasants who engage in subsistence agriculture. We do not know how technological advances and their various implementations, will help or hurt farmers. The inequalities perpetrated by the modern use of digital technology will mean that not all people will benefit. We are now more into virtual communication rather than cherishing what they have in real life.

Digital Urbanism provides immersive new world of ubiquitous connectivity, social media feeds, smart phones, mobile apps, responsive design, algorithmic recommendation systems etc. We are nowadays are addicted to mobile phones, and are now missing the right emotions, fun in real life. With so much content on the Internet, there are high chances of plagiarism and it is not easy to copyright everything. Studies consistently show that more social media causes loneliness, social anxiety, and depression. Moreover, wearable technology, tech implants, AI-medicine, autonomous robot workers and companions, and many other coming technologies will allow humans to reach new limits of what to do and expect. We do not know how many members of the society will be able to be part of the enjoyment of that ubiquitous, hyper-connected, AI- tech society. No product is made today, no person moves today, nothing is collected, analyzed or communicated without some digital technology being an integral part of it. We are living a digital life in every upgrading digital world. We all love technology so much as they have made our lives simple and easier, and have forgotten about pros and cons of digital life.

Digital Urbanism tries to integrate desperate systems to enable easier access to municipal services, provide sustainable environments, and foster innovation. People use data and digital technologies to deliver results that are more relevant and meaningful to residents. More comprehensive real-time data gives the ability to watch events as they unfold and respond with faster and lower-cost solutions. It also consists of specific applications to translate raw data into alert insight and action requiring the right tools and this is where technology providers and app developers come in. Tens of millions of people in cities worldwide begin and end everyday fuming in traffic or piling into overcrowded buses and trains. Smart mobility applications have the potential to cut commute time and improve the daily commute, critical to quality of life. Using digital signage or mobile apps helps deliver real-time information about delays, enables riders to adjust their routes. Deploying a range of applications, cities can help fight crime and improve other aspects of public safety. It is also about using technology and data purposefully to make better decisions and deliver a better quality of life.

Digital Urbanism injects technology more directly into the lives of residents: smart phones have become the keys to the city, putting instant information about transit, traffic, health services, safety alerts, and community news into millions of hands. With the increasing urbanization, cities need to make greater use of Artificial Intelligence-driven applications and systems. We live and work predominantly in urban environments, and to a great extent economic, innovation, and business activities are concentrated in cities. It has been unanimously assumed that the role of cities is essential to achieve community objectives to amore efficient use of resources. Digitization starts with people, not technology, as it is not just about installing digital interfaces in traditional infrastructure or streamlining city operations. Digitalization of cities need to improve some key quality-of-life, and translates into lives saved, fewer crime incidents, shorter commutes, a reduced health burden, and carbon emissions averted. Quality of life has many dimensions, from the air residents breathe to how safe they feel walking the streets. It is vital to understand citizen needs, and to design systems around citizen outcomes for the future.

Digital Urbanism must aim to be rather than a functional utilitarian city that worked more like a machine, to a living and breathing city, where people work, play, and produce. The city itself has to become very much part of their lives. There are increasing concerns that technology should support and not ruin the life of citizens. The talking point is how to strike the balance. We know the application of new technologies can change of the city life, but we need to find a balance. Young people want to live the experience, and they find it very easy to live in this environment, which is a generational thing. There has to be more sustainable development- embracing the green, social, and economic aspects of sustainability. The new digital revolution is based on knowledge and that destroys jobs and young people worry about finding good quality jobs. All cities face challenges to become sustainable, and these challenges are concerned mostly with environmental impact, and socioeconomic progress. The digital transformation of city has to make it liveable with smart mobility, and green lungs. With a significant rise in data leaks and security flaws, there is a need to protect security and privacy.

Digital Urbanism advocates for the need to rethink our everyday interactions with digital infrastructures, navigation technologies, and social media as we move through the world. Everyday millions of people turn to small handheld screens to search for their destinations.

New technologies and artificial intelligence will drive the new trend- some good and some bad- in the way they impact our lives. Critics have argued that digital media alienates users from space and place. ICT are proven as enablers of change and have a greater potential to continue promoting sustainable growth. The Digital City will be able to leverage new technologies much faster, be more open, and better engage citizens. It has to take full advantage of opportunities brought about by highly transformational data-driven technologies. Moreover, public connectivity and a large-scale civil digital infrastructure deployment will enable better learning and better digital skills for all citizens. As technology changes so fast these days, cities should shift their focus from the specifics of the technology to platform thinking as well as how they can make data available in a consistent, secure way for innovation.

Digital Urbanism is a transition from Analogue City to Digital City. While we identify urbanism with the physical entity of the city, our knowledge of the nature of urbanism, and the process of urbanization is meagre. Our political and social structure has been hitherto formed predominantly by the Analogue City- it reflected to varying degrees both the inheritance of print culture and the conditions created by electronic media. Some of us have inhabited both the Analogue City and Digital City, while an increasing number of us have known only the Digital City. We might say that our public spaces is now inhabited by the citizens of two cities- the Analogue City and the Digital City. Only recently has the Digital City begun to manifest itself in the public spaces that have been hitherto ordered by the Analogue City. The term Analogue City also refers to the phenomenon of the overhead and underground pedestrian connections. The collection of walkways and subways, flyovers and tunnels connecting railway stations, hotels, shopping malls, official blocks, and leisure centres allowed citizens to move about Urban Centres. New technologies not only alter the structure of our interests, but also the nature of community.

Digital Urbanism is a transition to a society based on virtual, intangible, vectors, using computing techniques and algorithms. The growth of Digital Cities and the Urbanization of the world is one of the most impressive facts of modern times. We are in the history of the Digital City- the Internet has been around for a half-century- the World Wide Web as the part of the Internet we access through Web Browsers is about thirty years old- and our identity with online presence began just over fifteen years ago. The transition to smart phones and tablets, which made digital media a constant presence in our lives and our default media environment has occurred only over the last decade. What we are witnessing is the ascendancy of the Digital City, which is characterized primarily by the advent of ubiquitous Internet connectivity, no longer just at home or work, but also on mobile technology. In a way our cities are already digital. It is actually the hidden stuff, such as data, that have the power to really change the lives of citizens. It tools are to improve the quality of data infrastructure, services delivery, and citizens interface in existing cities. Lots of cities are coming up with different initiatives.

Digital Urbanism during 1990s and 2000s has created Digital Cities, and become operational throughout the world. One of the first definition about the Digital City talks about a kind of city that is substantively an open, complex, and urban information resource, which forms a virtual digital space for city. Each country has undertaken a different technological development path towards digitalization with ITC playing a central role in most of them. Augmented and virtual reality of the city is another facet of exposing or simulating city data from the past, present, or future. The world is moving towards truly digital urban centres with urban population increasing to 6.3 billion by 2050, which is two thirds of the world population. While the Digital City mainly considers the technological features, the Smart City concerns all the aspects of everyday life. It seems natural to regard today's Smart Cities as the successor of Digital Cities. It is also natural to think that their differences are due to the technologies they use. Digital Cities are characterized by activities based on Web services, while Smart Cities demonstrate sensory services. Digital Cities explore cyberspace and Smart Cities exploit physical space.

Digital Urbanism has produced a large amount of literature on definitions and concepts of digital urban centres. Many scholars state that the Digital City emerged before the Smart City, and the Smart City is an evolution of the Digital City. With the substantial volume of activities under the name of Digital City, the conceptual relatives of it include Smart City, Intelligent City, Virtual City, Ubiquitous City, and Information City. Digital Cities provide innovative services based on broadband communication, and service-oriented computing, while Smart Cities apply technologies of self-monitoring and self-responsive systems to complex social problems. The technological aspect highlights the difference between Digital Cities with the rise of the Internet and Smart Cities being challenged in the era of Internet of Things. Of course, the next stage is to evolve networked society based on cyberspace physical systems. In addition, it is possible to observe that the Digital City is still today the most cited issue keeping its own relevance in spite of the development of the Smart City, which today is one of the hottest topics in the domain. A focus on the Smart City can be better performed from studies on the Digital City.

Digital Urbanism in public administration, and not least urban planning is gaining momentum. Urban planning is a matter of public interest. Digital urban planning is an innovative process of simulating the physical environment based on precise and up-to-date sensor-fed data in order to facilitate the decision-making process for all stakeholders. Digital urban planning is becoming a reality for a growing number of cities around the world. On the physical side, the perspective of composition, it is a dual of physical, and technical planning. On the physical side, the process includes the city infrastructure itself along with in spatial attributes, and functional departments of the city administration. On the technical side, it is the vast ICT infrastructure, distributed data bases, and a management system that ties all together, and interfaces with end users. On the process level, digital urban planning is a complex iterative process that constantly revolves around acquiring data from multiple sources (sensors, cameras, IoT devices, meters) and processing it properly (conversion, validation, summarization, aggregation), and turning it into meaningful insights that end users can access through an online platform.

Digital urbanism is creating significant opportunities for social and economic development and more sustainable living. Digital future refers to the idea that all businesses will operate digitally in the future. The digital future is ambient computing, wherein consumers are constantly connected, so in the future businesses will operate in the same way. It is already happening in banking and financial services, retail, telecoms and travel, every other business is building a digital future based on an always on connected interaction with every customer. All businesses recognize they must offer a personalized experience based on the actuality of the customer activity, history, status, and ultimately motives and aspirations. We have been simply adapting to our new digital reality because we had to. All of the digital technology has allowed our economy, our education, and our emotional selves to keep going in an impossible situation. Countless volunteers, activists, entrepreneurs, businesses, and governments make the Internet and the digital world it enables. All signs point to the fact that users continue to buy, carry, and wear smart devices in increasing numbers from phones to watches and more.

CHAPTER FOUR

Smart City

"Adding lanes to solve traffic congestion is like loosening your belt to solve obesity"- Glen Hemistra

Smart City is predominantly composed of Information and communication technologies to develop, deploy, and promote sustainable development practices to address growing urbanization challenges. Cloud-based Internet of Things applications receive, analyze, and manage data in real time to help municipalities, enterprises, and citizens make better decisions that improve quality of life. Pairing devices and data with a city's physical infrastructure and services can cut costs and improve sustainability. Communities can improve energy distribution, streamline trash collection, decrease traffic congestion, and even improve air quality. Municipal governments are leveraging on cellular wireless technologies and Internet of Things solutions to connect and improve infrastructure, efficiency, convenience, and quality of life for residents and visitors alike. Together, these smart city technologies are optimizing mobility, public service, and utilities in a non-ending phenomenon of urbanization. Secure wireless connectivity and Internet of Things technology are transforming traditional elements of city life. The advent of 5G technology could propel Smart City technology to higher levels.

Smart city uses information and communication technologies or ICT to increase operational efficiency, share information with the public, and improve the quality of services along with the welfare of citizens. The overall mission of a smart city is to optimize city functions and drive economic growth. It also improves quality of life for its citizens using smart technology and data analysis. There are several major characteristics of a smart city which include: a technology-based infrastructure, environmental initiatives, an efficient public transport system, a confident sense of urban planning, and people to live and work within the city and utilize resources. It is important to form a strong relationship between the government- bureaucracy- and private sector. This relationship enables and maintains a digital, and data-driven environment outside of the government. For example, surveillance equipment for busy streets could include sensors from one company, cameras from another, and a server from yet another. Therefore, a smart city's success becomes more focused on building positive relationships among all stakeholders in a collective manner.

Smart City technology consists of a combination of the Internet of Things devices, software solutions, user interfaces, and communication networks. They rely definitely on first and foremost on the Internet of Things. The Internet of Things is a network of connected devices- such as sensors or home appliances- that can communicate and exchange data. Data collected and delivered by the Internet of Things sensors and devices is stored in the Cloud or on servers. The connection of these devices and use of data analytics facilitates the convergence of the physical and digital city elements. This improves both public and private sector efficiency, enables economic benefits, and improves citizen's lives. The Internet of Things devices sometimes have processing capabilities called edge computing. Edge computing ensures that only the most important and relevant information is communicated over the communication network. A firewall security system, that is necessary protects, monitors, and controls network traffic within a computing system. Firewalls ensure that the data constantly being transmitted within a smart city network is secure by preventing unauthorized access to Internet of Things network or city data.

Smart city technologies include: application programming interfaces, artificial intelligence, cloud computing, dashboards, machine learning, machine to machine, mesh network etc. Emerging trends such as automation, machine learning, and the Internet of Things are driving smart city adoption. A classic example of a smart city initiative is the smart car parking meter that uses an application to help drivers find available parking spaces without prolonged circling of crowded city blocks. The smart meter also enables digital payment without any risk. Smart public transit is another facet of smart cities: smart traffic management is used to monitor and analyze traffic flows in order to optimize streetlights and prevent roadways from becoming too congested or overcrowded based on time of day or rush-hour schedules. Ride-sharing, and bike-sharing are also common services in a smart city now. Using smart sensors, smart streetlights dim when there are not cars or pedestrians on the roadways. Smart grid technology can be used to improve operations, maintain, and plan, and to supply power on demand and monitor energy outages.

Smart City initiatives also aim to monitor, and address environmental concerns such as climate change, and air pollution. Waste management and sanitation can also be improved with smart technology. The Internet connected trash cans and Internet of Things enabled fleet management systems can be used for waste collection and removal, or using sensors to measure water parameters. It can be used to guarantee the quality of drinking water at the front end of the system with proper wastewater removal and drainage at the backend. Smart City technology is increasingly used in improving public safety, from monitoring areas of high crime to improving emergency preparedness with sensors. Smart Sensors, for example can be critical components of an early warning system before droughts, floods, landslides, or hurricanes. Smart buildings constructed with sensors not only provide real time space management, but also ensure public safety. It can also be used to monitor the structural health of buildings, by detecting wear and tear, and notify repair needs. Sensors can also be used to detect leaks in water mains and other pipe systems, helping reduce costs, and improve the efficiency of public workers.

Smart City utilizes its web of connected Internet of Things devices and other technologies to achieve their goals of improving the quality of life, and economic growth. These technologies also bring efficiencies to urban manufacturing, energy efficiency, space management, and fresher goods for consumers. Smart technology will help cities sustain growth and improve citizen's welfare, while it could alleviate detrimental effects. With proper coordination, electric vehicles could also be used to regulate the frequency of the city's electric grid, when they are not in service. Autonomous vehicles or self-driving cars, could potentially change a population's perspective on the necessity of owning a cars. Smart city initiatives need to include the public, residents, and visitors. By this smart citizen gets Smart City applications enable cities to find and create new value from their existing infrastructure. It plans to make the data transparent and available to citizens, often through an open data portal or mobile app. Through a Smart City app, residents may also be able to complete personal chores: viewing their energy consumption, and paying bills.

Smart City yields a high quality of life to its residents, while also generating overall economic growth. A major advantage of a Smart City is its ability to facilitate an increased delivery services to citizens with less infrastructure and cost. Many cities across the world have started implementing smart technologies in order to facilitate new revenue streams, and operational efficiencies. For example, the city-state of Singapore uses sensors and Internet of Things-enabled cameras to monitor the cleanliness of public spaces, crowd density, and the movement of vehicles. Its smart technologies help companies and residents monitor energy use, waste production, and water use in real time. It has an elderly monitoring system to ensure the health and well-being of its senior citizens. For example, the city of Barcelona, Spain, has smart transportation systems, and smart bus systems, which are complemented by smart bus stops that provide free Wi-Fi, USB charging stations and bus schedule updates for riders. For example, the city of Dubai, United Arab Emirates uses smart city technology for traffic routing, parking, infrastructure planning, and transportation.

Smart City has four essential elements necessary for thriving: pervasive wireless connectivity, open data, security we can trust, and flexible monetization schemes. Evolving Low Power Wide Area Network technologies are well suited for most Smart City applications for their cost efficient and ubiquity. For example, Amsterdam is a well-connected Smart City reaping the rewards of opening the data vault. The city built autonomous delivery boats called reboats keep things moving in a timely fashion. It shares traffic and transportation data to interested parties such as developers, who then create mapping apps that connect to the city's transport systems. However, connected cameras, intelligent road systems, and public safety monitoring systems can provide an added layer of protection and emergency support to aid citizens when needed. The security objectives are availability, integrity, confidentiality, and accountability. Smart City needs to establish sustainable commerce models that facilitate all ecosystem players' success. Singapore has been ranked the world's smartest city for its Smart City Technology for enhancing sustainability and better serving humanity.

Smart City projects in Tokyo underpin improving energy security and efficiency; and showcasing advanced technology. They have introduced energy efficient solutions, including high efficiency systems, electric vehicles, and local power storage. Smart City technologies used in Tokyo is to develop into a Smart-Energy City. Urban design development of Tokyo has a goal to maintain the total amount of greenery in the city. As the global digital economy is expanding, cashless payments in Japan are making their way into everyday life all around the world with advantages of ability to simplify payment sending. Today we are seeing only a preview of what technology could eventually do in the urban environment. Cities are getting smarter and are becoming more liveable and more responsive. Cities are moving beyond the pilot stage, and using data and digital technologies to deliver results that are more relevant and meaningful to residents. Smart technologies are being injected more directly into the lives of residents. Smart phones have become the keys to the city transit, traffic, health services, safety alerts, and community news into millions of hands.

Smart City technology has three layers work: technology base, specific application, and usage by cities, companies and the public. The applications perform differently from city to city, depending on factors such as legacy infrastructure systems, and on baseline starting points. Applications can help cities fight crime and improve other aspects of public safety. Digital Twin technology has been around for decades and work like sophisticated 3-D maps, but ones hooked up to real-time data collected from the real world. Simulated cities can tap into huge amounts of data on things like traffic, people's movements, power systems, streetlights, and the weather. A digital twin platform is used by urban planners to test innovations. As the digital economy grows and matures, the Smart City movement is gaining momentum. For many people it represents the promise of high tech cities, with autonomous cars rolling on the streets, drones delivering food, and connected devices everywhere helping city dwellers to perform a myriad of activities. A city only becomes truly smart, when all citizens are ready for it. Building a smart workforce is another aspect of ensuring that Smart City initiatives are adopted.

Smart City is about data, a lot of it- the digital shadow of a city is going to process a huge amount of information on a regular basis. Internet of Things tech ties everything together, mainly via sensors, enabling the gathering of information which will serve everybody to make the city function better. Internet of Things devices such as connected sensors, lights, and meter can collect and analyze data for the better functioning of Smart City. It is a system of interconnected computing devices, machines (mechanical and digital), and objects provided with unique identifiers, and the possibility for each member to transfer data over a network. From here, the city can use this data to improve infrastructure, public utilities, services, etc. In other words, Internet of Things is a real-time communication tool that will permit us to access information. Data exchange is essential for a Smart City to succeed in their promise to deliver safer, healthier, and more sustainable communities. A well-planned infrastructure transforms cities into vibrant socio-economic communities. Proper data governance should regulate how data is collected, handled, and shared.

Smart City is about innovation enabled by new technologies, to address not only technical aspects, but also the needs of partners, citizens, and institutions. Smart City has Closed- Circuit Television (CCTV) Cameras, which serve many purposes, including crime prevention, traffic monitoring, weather prediction, as well as observing and monitoring several areas. However studies prove that more cameras do not necessarily amount to a lower number of crimes or more cameras do not necessarily amount to a higher degree of perceived safety. Though this could be instrumental in keeping our city safe, it needs to be supported by efforts to strengthen other aspects of safety and security. China has the most number of CCTV cameras in use. Network infrastructure is critical to realizing the potential, defined by its high-speed, low-latency data transmission, and ubiquitous connectivity. A new telecom technology utilizes machine learning combined with signal processing algorithms will allow video transmission from all connected cameras over the cellular network in an optimal and robust way. Robotic process automation and artificial intelligence are clearly in the core of future cities.

Smart City needs new and advanced solutions to meet the security challenges of the 21stcentury. It aims to have an enabling platform that delivers advanced services for businesses, and residents, helping to bridge the digital divide, and provide a better quality of life for all. Smart LED Streetlights are connected by a Wi-Fi signal, so that they are remotely controlled and last longer reducing maintaining casts, and save energy. Soon, there will be more intelligent Streetlights, which will oversee parking spots, check for illegal activity, or even measure air humidity- thanks to built in sensors, cameras, microphones, Wi-Fi, and Bluetooth. Internet of Things for the average person is the Smart Phone, because it is going to be everywhere and everyone carries it all day. We use it for a large number of daily tasks to interact with other smart devices. We interact with Internet of Things using a Smart Phone simply because this is the computing platform that we are most likely to have with us at any point of time. Consumers will have more options, when it comes to smart devices that means interaction with those devices is through the Smart Phone. Survival of intelligent cities depends on its sustainability.

Smart City technology relies mainly on Internet of Things to create an information-driven, connected base. The Cloud-based nature of Internet of Things solutions is appropriate for any Smart City by sharing a platform based on open data. The development of a Smart City requires the storage capacity, scalability, security, and data processing of users and enterprises. A Cloud functions as a storage and analysis system for the data used in everything connected in a Smart City. Personal Computers and Server Files, Web Page meta-data, images, and video, and data created by machine-to machine communication will all be housed in the Cloud. All innovations are to be centred on the citizen, adhering to the principle of universal design and usable by people of all ages, and abilities as a means to improve the lives of the city inhabitants. Though people continue to be drawn to cities by the economic, social, and creative opportunities they offer, large cities are more productive than rural areas, producing more patents and yielding higher returns on capital. Smart City is a model to manage its resources in a new and more efficient way that relies largely on technology.

Smart City systematically applies digital technologies to reduce resource input, improve citizen's quality of life, and increase the competitiveness of the regional economy sustainably. 5G- is a key technology when it comes to digital transformation. 5G- wireless technology or 5th generation mobile network is meant to deliver higher multi-Gbps peak data speeds, ultra low latency, more reliability, massive network capacity, increased availability, and a more uniform user experience to more users. 1G-delivered analogue voice, 2G- introduced digital voice, 3G- brought mobile data, and 4G- ushered in the era of mobile broadband. 5G- technology will permit the interconnection of massively connected objects in Internet of Things with data transmission speed several times faster. With high speeds, superior reliability, and negligible latency, 5G- will expand the mobile ecosystem into new realms. Also, all major Android phone manufacturers are commercializing 5G-phones. We are likely to the development of increasingly sophisticated mobile phones, and phone apps that take advantage of faster download speeds and stronger broadband connections.

Smart City innovations are likely to have: parking apps that show drivers where the nearest available parking –spot; all-digital and easy-to-use parking payment systems; city guide app with information about museums, parks, landmarks, public art; restaurants and real time traffic data; Wi-Fi in subway stations and on trains, along with weather information at every station; sustainable, and energy efficient residential and commercial real estate; dynamic kiosks that display real-time information; concerning traffic, weather, and local news; app or social media-based emergency alert and crisis response systems; broadband internet access for all citizens; mobile payments everywhere food, apparel, and public transport, etc. However, urbanization also presents major challenges. The world's fastest growing cities have seen problems adjusting to growth and industrialization, choking under the burden of pollution, congestion, and urban poverty. Urban settings magnify global threats such as climate change, water and food security, and resource shortages, but also provide a frame-work for addressing them. Small-scale projects such as cycle lanes, bike sharing, and planting of trees may help reduce climate change.

Smart City initiatives are typically complex by nature, involving many stakeholders and mobilizing significant resources. New technologies alone without any social impact do not actually have value. Plans for more wired, networked, and connected urban areas face challenges. Some of the drawbacks include security issues in terms of public data, digital equity, personal privacy, excess network trust, cyber hacking, and lack of social responsibility. These disadvantages of Smart City will help in the successful implementation of challenging factors, which can come in between the growth of the Smart City. The consistent barriers to Smart City include siloed, piecemeal implementations, growing expectations, and access to city services, uninformed citizens, engaging the public, shrinking budgets, and investment capital. Citizen's awareness of Smart City is remarkably low, where illiteracy is already a major issue. The dark side is the rapid urbanization within cities compared to its growth. The Phrase 'go digital, or go home' is very much here to stay. The heralded rise of Smart Cities is expected to bring data-centric solutions to urban challenges.

Smart City is definitely the promising future for a high quality communal life. Various cities throughout the world have started their efforts in designing, and implementing Smart Cities. Meeting citizen's requirements economically and effectively is the most important objective of Smart Cities. Research, development, and production in various relevant technology fields is accelerating. Remotely piloted aircraft systems are being widely studied, and developed due to their mission flexibility, recognizable architecture, and low cost. Drones are designed to replace road deliveries, so as to overcome infrastructure challenges, as it consumes less fuel, and consequently have a smaller impact on the environment. Providing users with awareness and control about privacy-sensitive information flows is a major challenge in Internet of Things scenarios. Fog computing technology for Internet of Things applications provide low latency supports to mobility, location awareness, scalability, and efficient integration with other systems such as Cloud computing. The technology of wireless sensor networks have evolved leading to sensors with increased memory, storage, processing, and communication capability. The use of unmanned aerial systems is emerging more and more in recent years.

Smart City is defined as a city in which ICT is merged with traditional infrastructures, coordinated, and integrated using new digital technologies. Technology trends allow us to create spaces in which humans and technology interact in a more connected, intelligent, and automated way. The critical components of the Smart City of the future are simple Artificial Intelligence, Big Data, and Cloud combined with 5G as the transportation medium. The vast speed of Wi-Fi connectivity, Internet of Things, and CCTV Cameras are able to harness technology to improve resident safety, and boost incident respond time. One of the really exciting things is apps that allow citizens to report local issues more easily or community networking platforms that allow neighbours to connect, and share resources. The digital transformation of a Smart City today would follow the concept of Device-Pipe-Cloud application. Devices collect data; the pipe delivers the data over the transportation infrastructure to the Cloud; where it is stored, classified, and cross correlated, and made available for processing by applications for visualization to make informed decisions. The Cloud is central to a successful Smart City-one of the powerful additions.

Smart City has an effective strategy that involves: defining the relevant Smart City concepts; designing the planning process; engaging and drafting approaches with stakeholders; as well as prioritizing initiatives; and crafting the road map. Budgeting limitations often constrain the pace at which cities can realize their development potential. A city only becomes truly Smart City, when all citizens are ready for it. Urban planners and innovators might develop personas of the ideal Smart Citizen, as they prepare for true plans for their cities. This often assumes that citizens enjoy Internet access, and are tech-savvy enough to use and interact with the city's spaces and services. It is equally important to provide accessibility to both the Internet and the devices to utilities, online capabilities, as well as having the technological skills to utilize the capabilities of a Smart City. Building a smart workforce is another aspect of ensuring that Smart City initiatives are adopted. A holistic strategy ensures that all ages have access to technology, education, and the opportunities to add value and have a part to play in developing the city. It is important to have the technological skills for effective use with open data policies.

Smart City development requires investment in reliable technology, and high speed connectivity. The accelerated development of new technologies including 5G, Ai, Cloud, and Edge Computing will help drive the evolution of Smart city. Edge computing is critical support the exponential increase in the number of connected devices, and vast growth in data collected. As vital infrastructures become connected, cities must be aware of vulnerabilities to adversaries. Telecom and technology companies must increasingly collaborate with governments, and invest in reliable networks, cyber security, and backup systems. Decarbonizing the sector is one of the most cost-effective ways to mitigate climate change. Government policies, teamed with financial incentives for companies to invest in smart buildings, are crucial to help transition toward accessing major energy savings, whilst improving energy services. There is immense pressure to transition to lower-carbon energy systems. Access to clean water and the ability to treat waste water are growing concerns for cities. On the waste side, using sensors can optimize the collections, and emphasize .distributed waste-to-energy solutions.

Smart City and Smart Growth largely overlap, and address multiple environmental considerations, enjoy wide currency- most urban planning is now based on these principles. Urban planning gets its start as a profession largely dedicate to averting different types of crises arising from urban growth, and provide conditions for public health. Planners who promote Smart City seek to develop a stronger sense of place through a more compact way of development, also known as mixed-use. Mixed-use development combines residential areas with places of employment, and commerce, instead of isolating individual areas, allowing for more pedestrians and public transit as opposed to traffic and congestion. The most important technological constituents of Smart City planning process include: 3D- visualization, including AR and VR; cloud-based services that aggregate, process, and store relevant data; user-facing portals, and mobile applications, cyber security tools; Internet of Things infrastructure capturing and feeding data to multiple destinations, advanced GIS services, big data management, and analysis tools. A Smart City is built on technology, but focused on outcomes.s

Smart City needs to preserve urban forestry as they are real value to communities. Among their many benefits, trees reduce energy costs, intercept air pollutants, store carbon, and reduce storm-water runoff. Urban forests can provide many important ecological functions and economic benefits, but continuous delivery of those services depends on the long-term health and resilience of the population. The benefits are greater with the tree canopy cover per capita. In urban settings, a tree's ability to intercept or slow rainfall reaching the ground can reduce the amount of overflowing onto paved surfaces and lost as storm-water runoff. Species selection makes a difference, and the difference tends to increase as trees grow, as they are vulnerable to climate change. Urban green infrastructure, including urban forests, is an important strategy for providing public goods and increasing resiliency, while reducing ecological footprints and social inequity in Smart Cities. Trees give roads a breath of fresh air, and help us feel better. Strategically planting trees near busy roadways may significantly enhance air quality. Selecting and locating trees can maximize climate, energy, and environmental benefits.

CHAPTER FIVE

New Urbanism

"People make cities, and it is to them, not buildings, that we must fit our plans"- Jane Jacobs

New Urbanism has emerged over the past two decades as a controversial alternative to conventional patterns of urban development. New Urbanism has a central theme that many activities of daily living should occur within walking distance, allowing independence to those who do not drive, especially the elderly, and the young. New Urbanism wants interconnected networks of streets to be designed to encourage walking, reduce the number, and length of automobile trip, and conserve energy. New Urbanism plans a broad range of housing types, and price levels to bring people of diverge ages, races, and incomes into daily interaction, strengthening the personal and civic bonds essential to authentic community. New Urbanism desires civic, institutional, and commercial activity to be embedded in neighbourhoods not isolated in remote areas, and schools to be sized and located to enable children to walk or bicycle to them. New Urbanism distributes green parks and community gardens within neighbourhoods. New Urbanism seamlessly links urban architecture and landscape to their surroundings as transcends style.

New Urbanism, an umbrella term, which encompasses neo-traditional development as well as traditional neighbourhood design, lives by an unswerving belief in the ability of the built environment to create a sense of community. New Urbanism seeks to foster place identity, sense of community, and environmental sustainability. New Urbanism emphasizes not only on the local and the particular, nostalgia for the past, historical traditions, but also the need for environmental care. New Urbanism encourages the development of planned communities in which people can live, shop, work, go to school, worship, and recreate without having to travel great distances by automobiles. New Urbanism is an urban design movement to create pedestrian-oriented settlements that also advance social equity, and mitigate the environmental impacts of development. New Urbanism has a policy framework that promotes an urban development pattern characterized by high population density, walkable, and likeable neighbourhoods, preserved green spaces, mixed-use development, available mass transit, and limited road construction.

New Urbanism has a planned community features such as townhouses, single-family homes, and a conveniently located town centre with a grocery store, nice restaurants, good shops, a theatre, a dry cleaner, common areas, offices, healthcare services, a farmers market, a day-care centre, an elementary school, and a worship place. New Urbanism seeks to turn existing communities, and neighbourhoods into diverse districts cleaning up polluted and dilapidated areas. New Urbanism focuses on human-scaled urban design- more foot traffic spending less on cars and gasoline; smaller business spaces and parking lots; healthier lifestyle due to more walking; nearer to healthier restaurants; less incentive to sprawl; removal of restrictive and in correct zoning codes; smart green transportation; bicycles as the most sustainable form of transport; and provides daily exercise for riders. New Urbanism is a planning and development approach based on the principle of how cities have been built for the last centuries: walkable blocks and streets, housing and shopping in close proximity, and accessible public spaces. New Urbanism is holistic with all built environment work together to create great places.

New Urbanism creates places that enrich, uplift, and inspire the human spirit- taken together, which add up to a high quality of life well worth living. New Urbanism provides higher quality of life, better places to live, work, and play; higher and more stable property values; less traffic congestion and less driving; healthier lifestyle with more walking; and less stress; close proximity to bike trails; parks and nature; pedestrian-friendly communities; friendlier society; more open space to enjoy; more efficient use of tax money with less spending on spread out utilities and roads. New Urbanism goes for green transportation: a network of higher quality trains connecting cities, towns, and neighbourhoods together; pedestrian-friendly design that encourages a greater use of bicycles, scooters, and walking as daily transportation. New Urbanism respects ecology and value of natural systems with energy efficiency, less use of finite fuels, more local production, and more walking, and less driving. New Urbanism focuses on the physical design of communities to create liveable, and walkable neighbourhoods.

New Urbanism focuses on design, which is critical to the function of communities. New Urbanism may transform deteriorating public housing into liveable mixed-income neighbourhoods. New Urbanism aims to provide plazas, squares, sidewalks, cafes, and porches to host daily interaction, and public life. New Urbanism streets are designed for people-rather than just cars- and accommodate multimodal transportation including walking, bicycling, transit use, and driving. It combines appropriate design elements that make places that are greater than the sum of their parts. Commercial strips with single-use development, and excessive asphalt are transformed into lively main streets or boulevards through new urban design. New Urbanism creates the walkable, vibrant, beautiful places to work better for businesses, local governments, and their residents. New Urbanism is a design movement towards complete, compact, and connected communities, particularly rich source for directed planning and development in recent years. New Urbanism creates urban design codes to physically define streets, and public spaces of shared use.

New Urbanism stands for the restoration of existing urban centres within coherent metropolitan regions, the reconfiguration of sprawling suburbs into communities of real neighbourhoods, and the conservation of natural environments. New Urbanism advocates that neighbourhoods to be diverse in use and population; cities to be shaped by physically defined and universally accessible public spaces; and citizen-based participatory planning and design. New Urbanism has the axiom that farmland and nature are as important to the metropolis as the garden is to the house, and so development patterns should not blur or eradicate the edges of the metropolis. New Urbanism brings cities into proximity, and a broad spectrum of public and private uses to support a regional economy that benefits people of all communities. New Urbanism pleads for transit, pedestrian, and bicycle systems should maximize access and mobility throughout the region while reducing upon automobile. New Urbanism envisages the promotion of rational co-ordination of transportation, recreation, public service, housing, and community institutions.

New Urbanism configures streets and squares to be safe, comfortable, and interesting to the pedestrian in order to encourage walking, and enables neighbours to know each other and protect their community. New Urbanism prefers natural methods of heating and cooling as more resource-efficient than mechanical systems. New Urbanism opposes the present fragmented sprawl- including roads, buildings, finance, and design- are geared towards conventional suburban design. New Urbanism has the idea to create something that is not only a source of agreement among urbanites, but something that has the power to allow other people to use them and carry on with remaking the physical world. New Urbanism offers up a system, where people are given back their stolen time, making the world see a new possibility for how we all spend our time and energy. New Urbanism rests on the understanding that great attention to detail is necessary to sustain peoples' interest when they are moving at a more leisurely pace. New Urbanism manifests itself inclusively in various ways so that urban settlements bring people together, not drive them apart.

New Urbanism is a complex planning paradigm and social moment that has recently become influential in planning. The charter of the New Urbanism and its movement has spread from the United States to other countries. Formal order, discipline, and hierarchy as well as grand, universal styles, solutions and formulas are to become recipes for the betterment of humanity with the world to become predictable and safe. But its opponents claim that the proper design of space leads to the development of a local community. They feel that New Urbanism approach is often one-sided, presenting either its advantages or disadvantages. Postmodernism rejects simple hierarchies and dichotomies, and emphasizes multiplicity and difference, not unity. The proponents of New Urbanism claim that present and future problems might be solved by drawing inspirations from the past. Numerous research results show the relation between physical elements of people's environment strengthens the ties among the inhabitants. New Urbanism ideas are a direct reaction to such problems as spatial segregation of inhabitants, concentration of poverty, excessive dependence on cars, and spaced out houses.

New Urbanism views the decentralized, auto-oriented suburb a recipe for disaster. New Urbanism blames these suburbs for ever-increasing congestion on arterial roads, a lack of meaningful civic life, the loss of open space, limited opportunities for people without cars, a general discontent among suburbanites. New Urbanism owes much to the city beautiful, and garden city movements of early twentieth century. New Urbanism focuses on a community's physical infrastructure in the belief that community design can help create or influence particular social patterns. New Urbanism claims to be committed to the concepts of strong citizen participation, affordable housing, and social and economic diversity. Many critics believe that, while New Urbanism contains many attractive ideas, it may have difficulty dealing with a wide range of contemporary issues such as scale, transportation, planning and codes, regionalism, and marketing. At the same time, it is important to appreciate the power of New Urbanism as an idea- it promotes a positive image of town life that includes the public as well as the private realm. New Urbanism requires a highly sophisticated effort in the remaking of a city.

New Urbanism rejects the key design tenets of modernist planning, and strives to revive pre-modernist planning, and strives to revive pre-modern urban forms. It contradicts the post-modern thought of commitment to pluralism. New Urbanism provides greater access to healthy foods, health services, and active lifestyles. New Urbanism increases social capital through greater social integration, supports culture, reduces household costs, stimulates commercial activities, and enhances quality of life. New Urbanism creates live/walk/work communities, and provides greater access to services and goods. New Urbanism decreases per capita infrastructure, utility, and facility costs, and encourages more efficient land use. New Urbanism increases street and pedestrian safety, reduces pollution, creates a cleaner environment, promotes proper waste management, and utilizes renewable energy. New Urbanism lays emphasis on beauty, aesthetics, human comfort, and creating a sense of place, and; special placement of civic uses within community. New Urbanism offers minimal environmental impact of development and its operation with eco-friendly technologies.

New Urbanism, initially conceived as an anti-sprawl reform movement, has evolved into a new paradigm in urban design. Over time, the planning and design concepts of New Urbanism have gained wider popularity, and have become diffused into development trends, and considerably influenced public policy. New Urbanism designers have contributed to the formulation and promotion of form-based zoning codes that focus more on physical form and less on land use to regulate new development. Advocates of New Urbanism have emphasized the role of physical design in addressing a number of socio-spatial problems from the initial stage of its conceptualization and diffusion. Some researchers have argued that most of these new Urbanism projects cater to high-income households, who self-select themselves into these neighbourhoods. New Urbanism type projects face considerable regulatory and non-regulatory barriers. New Urbanism requires a retrospective view of the problems and circumstances of urban and suburban development that contributes to its innovation as an anti-sprawl movement. Demographic changes pose new challenges for the future of New Urbanism.

CHAPTER SIX

Urban Science

Metro Sapiens: an Urban Species

Urban Science uses the city as a living laboratory to explore solutions with a meaningful purpose. This is the century of cities, and the age of the Metro Sapiens. Cities are drivers of change in the world. Cities are very much the dominant habitat of our human species. Cities have the potential to create sustained economic prosperity, and improve the quality of life for all. Cities are the epicentre of the world for cultural, art, and technology. New cities are a reality in the urban future of many nations. As the world population is increasing a rising number of people are moving from rural areas to more urban settings-this global phenomenon is called urbanization. Urbanization has been climbing steadily of late, with more than half of the world's population is now living in cities. By 2050, 70 percent of the world population is projected to live in urban areas. It is surprisingly difficult to define what exactly makes a city, or where a given city's boundaries lie. The rapid expansion of cities means that we are able to mass produce high-rise buildings that could fill the whole landscape. Design of our built in environment shapes our lives in both conscious, and unconscious dimensions.

Urban Science seeks to understand the fundamental processes that drive, shape, and sustain cities and urbanization. Our commitment to a global urban future demands that we move toward more sustainable urbanism. The processes that create cities, and the functions that go in cities arrange their internal makeup in particular ways. The most famous internal structure of cities is the concentric ring model with the centre of the city as the Central Business District or Downtown. Downtown signifies a city's greater brand, or identity, and has evolved into vibrant city centre. Downtowns are leading economic drivers for their cities, and highly inclusive places, given their access to opportunities and essential services for all users. Downtowns have intrinsic cultural significance with recreation, and entertainment opportunities. But, Intensive urbanization is swallowing municipal green areas, which causes intensification of erosion, decreases in biodiversity, and permanent fragmentation of habitats. Sustainable urbanism is society-based, complexity-led, and landscape-driven to shape a harmonious city. Cities are complex organisms with many stakeholders, as well as, a diverse population.

Urban Science aims to make cities more sustainable, resilient, and liveable. To live sustainably, humans need to embrace their inner urbanites- and recognize our species not as Homo Sapiens, but Metro Sapiens. The Urbanization of the world continues apace, and is one bright spot in an otherwise challenging global economic environment. Globalization and capitalism probably contributes to the creation of generic, highly consumptive human settlements, with poor regard for environmental consequences. Nature needs to be protected, and integrated into all aspects of urban life to provide the full range of diverse life supporting, and life enhancing benefits. Existing cities need to change as they grow to maximize liveability, minimize energy use, and environmental impact. Emerging cities can adopt technologies, policies, and designs to lower carbon footprints. Cities attract creative, and innovative individuals, and foster creative, engaged communities vie education, culture, and the space for interactions and discussions in order to emerge as vibrant centres of innovation. World City Reports reinforces the benefits of cities that engage all stakeholders and appreciates the value of sustainability.

Urban Science unfolds the common processes that influence the structure and dynamics of all cities. The United Nations General Assembly has designated the 31stOctober as World Cities Day. Metro Sapiens needs to structure the city in ways to be supportive of the natural environment. The concept of Smart City is relatively new and can be seen as a successor of information city, digital city, and sustainable city. Cities will be driven by smart technology that will make people connected in more ways than before, and hopefully boost city life experience. A Smart City is an urban centre that hosts a wide range of digital technology across its ecosystem. Smart Cities use technology to better populations living experiences, operating as one big data-driven ecosystem. The Smart City uses that data from people, vehicles, buildings etc to not only improve citizens' lives, but also minimize the environmental impact of the city itself. Sustainable cities of the future will redefine the phrase 'concrete jungle', becoming more of a literal expression than a symbolic one. The integration of smart technology and collaboration in our urban living environments has costs as well as benefits.

Urban Science depends heavily on upon digital technologies, which provide not only previously unconceivable analytical power, but also access to huge amounts of data. Many city dwellers around the world are connected to an entire ecosystem of apps, instant communication, video, and online deliveries. When it comes to Smart City living improvements in citizens' experiences can be seen across the board, from health factors to security, and day-to-day convenience. New Urbanism is a design movement toward complete, compact, and connected communities. New Urbanism is a technique for designing our built environment, and reforming it in such a way that it leads to a better standard of living for us, and improve the quality of our life in the places we live in. It primarily focuses on reducing the urban sprawl, which arises because of the low rise development in some areas. New Urbanism is about assembling all the basic facilities that may be needed by an individual at a particular place so that there is no physical separation. These contain housing, shops, work places, entertainment, schools, parks, and civic facilities essential to the daily lives of the residents within easy walking distance.

Urban Science provides a new and exciting focus on new cities rather as unexplored phenomenon. Sustainable cities of the future will be bigger, greener, and more intertwined with technology. The popular buzz words like smart technology, organic sourcing, and emphasis on natural resources will become the prominent themes of living in future cities. Cities will be more densely packed than ever before and there will be a need to create more localized communities. Metro sapiens in future will become sustainable urban habitat. Cities are largely unpredictable, because they are complex systems that are more like organisms than machines. Neither the laws of economics nor the laws of mechanics apply. The variously labelled cities of the future would need to be dynamic, and intelligent in every aspect of social, economic, and environmental sustainability. Cities of the future are likely to be a socially diverse environment, where communities are focussed around neighbourhoods. Internet of Things and open data will have severe impact on the experience, interaction, and wellbeing of citizens in future Smart Cities.

Urban Science pursues deeply quantitative and computational approaches to understanding the Smart City. Smart Cities harness the power of data from sensors in order to understand, and manage city system. There will be significant advancements of intelligent building design as a key constituent of eco-city development for creating greener, and effective built environments. Energy efficiency will become central challenge for urban life similar to energy efficient structuring of living organisms. Urban farms or roof-top gardens help reduce urban heat islands. The inevitable upcoming technology of autonomous vehicles will affect our cities, and several aspects of our lives. We have to plan and design a city of the future, educated by lessons of the past, and anticipating challenges of the future. Urban planning will be human-centric to create flourishing cities of the future. The kind of cityscape has to encourage walking, and biking as well as micro-communities, where people live, work, and play in the same vicinity. Smart buildings may incorporate natural elements, and are largely modular, leading to faster production with less waste.

Urban Science is emerging now as a coherent body of theory and knowledge that can contribute to a more sustainable urban world. Sky gardens interspersed green spaces will promote natural airflow in buildings, while providing shade, and social areas. Solar panels and roof gardens will encourage sustainable energy, and small-scale farming. Rainwater cleansing might collect and filter rainwater for reuse. All parks and infrastructure allow water to percolate through soil to recharge the water table. Lighter and cheaper bladeless wind turbines on rooftops can provide supplementary energy. Remotely programmed drones become large and powerful enough to transport people within the city. Future city for all are fully accessible to the disabled, giving all residents unfettered access to goods and services. Future of planning system for the 21^{st}century need to de-risk, and prove the demand for the data-driven, and digitally-enabled products and services. Urban development seeks to promote innovation, healthy living, and sustainability. The core principle behind an urban regeneration strategy could be massive mixed-use developments with a variety of housing types to encourage a diverse population mix.

Urban Science focuses on the metabolism of cities-studying them as ecosystems, characterized by stocks and flows of resources, including energy, water, capital, and information. The principles of Blue Zones can be applied to urban design to promote healthy living and longevity through a variety of design practices. Incorporating Eco-District sustainability practices into design can reduce carbon footprint and environmental impact. Cities need to play a key role in relation to low-carbon resilient future. There is no doubt that the cities in the future will strive to promote various aspects of sustainability more earnestly than today and there will be more technology at hand to be applied to achieve that end. Smart Cities are not just a buzz word, but a reality that is only getting bigger. In reality smart means adopting a more efficient management of services and turning cities into enablers of innovation, economic growth, and well-being. An increasing number of cities and metropolitan areas in the world are embracing the Smart City concept. Public-private partnerships, in fact, are critical to the Smart City. Information technology may be the main infrastructure and driver of the Smart City.

Urban Science is multi-disciplinary and draws theoretical ideas from across architects, engineers, ecologists, social scientists, and computer scientists. Green architecture plays a key role by creating a better quality of life. If people take pleasure in their surroundings, this contributes to higher productivity, and a reduction in energy consumption in a Smart City. Vertical greenery will be the defining feature of future metropolises as people integrate greenery into building's façade. This offers a practical spatial solution to environmental issues like carbon dioxide emissions, a by-product of urbanization. Green walks, roofs, and gardens offer protection from ultraviolet rays, reduces temperatures, and keeps building interiors cool, which in turn decreases demand for air conditioning, and curtails a building's carbon footprint. The high-rise towers featuring green walls can recycle their own energies, solar energy generators can churn out bio-fuel, and wind turbines can produce electricity. Indoor farming or growing crops is also a key element of vertical design. Digital technology has provided society with hundreds of advances that make life easier and better for everyone.

Urban Science analyzes available big data from such things as smart phones, Wi-Fi connected sensors, and satellites. From the Personal Computer to the Internet to the smart phone to the Internet of Things, we are increasingly living in a technologically driven world. Citizens can engage with Smart City systems using smart phones, and other mobile devices, including cars and homes. Being able to connect to a city's physical infrastructure, and services has the potential to cut costs, and improve the city's sustainability. Cities can improve energy dispersal, streamline city services, decrease traffic, and reduce air pollution. Wi-Fi hotspots on a larger scale can transform the way users access information. Increased use of public transportation reduces the number of cars on the road, parking needs will decrease, and enable cities to re-purpose land for housing. Well- developed smart technology tools cab help government agencies, the environment, and residents. The very definition of urban infrastructure is changing from mere physical assets to embrace data and technology. The jump in quality from 4G to 5G will be gigantic, and more momentous, while it will offer vast improvements to the smart phone.

Urban Science tries to enhance soft-ware enabled technologies and urban big data that have become essential to the functioning of cities. The vision for future cities, therefore, must find both scientific and democratic process, aligning the interests of all stakeholders. The main takeaway is that it is not just the physical infrastructure that needs to be emphasized, but also liveability. It is quite the challenge for city planners to turn cities into efficient and productive hubs. This involves everything from air quality to temperature, and sanitation as well as green areas such as gardens, and parks, and green infrastructure. Cities of the future will create not only value, but also comfort, fuelling an enduring way of life across the four domains of economics, politics, culture, and ecology- all need to be sustainable. The humble bike and electric car get all the talk, and they absolutely have an important role to play in future cities. The future cities energy will ideally come from both small-scale local projects, and large wind and solar farms-but it will be renewable and clean. The challenges in building a smart, efficient, sustainable, healthy, and liveable city are manifold.

Urban Science aims to make urban operational governance, and city services to become highly responsive to a data-driven urbanism, and helps improve city system or service to be managed in real-time. A vast deluge of real-time, fine-grained contextual and actionable data include: utility companies of electricity, water, and gas; transport providers; mobile phone operators; travel and accommodation web sites; social media sites; crowd sourcing and citizen science of weather, maps, and local knowledge; government bodies; financial institutions; public administration; private surveillance and security; emergency services; home appliances; and entertainment systems. Urban data has emerged as an excellent stream of constant, real-time, and accurate information about all urban activities. Citizens, who are both the cause and effect of the socio-economic dynamics of our society, are to be involved in all planning efforts. Future cities will need to be optimized across multiple scales. Future cities will need to be both soft and hard, and also to do more with less, while figure out ways to use and reuse resources widely.

Urban Science provides understanding of the relations between societies, environments, and specially mass and energy flows with population growth factor. Cities are complex natural ecosystems, even if we sometimes think human-made structures are not nature. Urban systems and cities are complex entities with their own metabolism, and the use of metabolism concept expands to include aspects of liveability. Urban metabolism is a concept in which the city is using the biological notion referring to the internal processes by which living organisms maintain a continuous exchange of matter and energy with their environment to enable operation, growth, and reproduction. Monitoring resource inflows and outflows is crucial for urban sustainability. Cities are natural ecosystems, obeying to the same laws as all other species. Urban metabolism concerns the many and varied interactions between societies and the biosphere. Urban ecosystem requires optimization of resources use in all sectors and levels. Urban areas are significant consumers of matter and energy, either directly or indirectly through the materials, goods, and services, they import or export- a sustainable development issue.

Urban Science studies will help suggest clear strategies in designing grey-green-blue infrastructure and more effective ecological services. Urban systems are dominated by humans to a greater extent even than agricultural land. Buildings are again parts of the wider urban organization, which plays a major role in just about all aspects of life, including health, water cycle, transportation, and what are called ecosystem services. Grey infrastructure refers to concrete drainage, water treatment, and transportation systems that provide little support for biodiversity. Green infrastructure denotes ecological framework at different levels, including more open terrace to absorb water, along with wetlands, botanical gardens, and other features more sympathy with nature. Blue infrastructure defines water supply and drainage, which should be coupled with green infrastructure. Green and Blue will always combine with Grey elements, but the profile and role of these spaces remains an open question in urban planning. Biogeochemical cycles of nutrients such as carbon, nitrogen, and phosphorus, or the flow of energy through food systems are a key focus of ecosystem ecology.

Urban Science critically reviews the emergence of complex interactions among human, natural, and technological systems and the uncertain trajectories that characterize urban future. Cities are now home to most of the world's population, generate over 90 percent of global economy, produce up to 75 percent of greenhouse gases, and consume 75 percent of energy, and 60 percent of global drinking water. It is increasingly evident that cities amplify the consequences associated with globalization such as movements of people and products, access to, and disruption of natural resources, and threats to biodiversity. The Physical configurations of urban settlements are also evolving with social and technological accelerations, which promote dissolution of boundaries among areas traditionally labelled as urban, regional, suburban, and rural. Convergence and interactions of functional, structural, and social changes result in challenges of unprecedented complexity for city governments. There is an effort to evolve a scientific theory of human settlements and their functions that is grounded in empirical evidence of how the many dimensions interact with one another.

Author Bio

Prof.RVM. Chokkalingam, a former lecturer/curator/scientist, now @ 77, is a Bangalore-based Science Writer. He has made lifetime contribution towards Public Engagement with Science for over 50 years. He is a Science Museum Scholar, specialised in the design of science exhibits from London Science Museum. He is a proponent of New Urbanism movement, and his special interest includes urban studies of nature-guided urban design, and data driven smart cities. He has published more than 160 articles in newspapers and magazines, and authored more than a dozen books. He is a recipient of Karnataka State Award of Science Communication in 2012. He is a paper airplane hobbyist.

www.ingramcontent.com/pod-product-compliance
Lightning Source LLC
Chambersburg PA
CBHW030913180526
45163CB00004B/1818